Passing the Minnesota MCA-II/GRAD Component Reading Test

Based on October 2006 Standards

Mike Kabel
Dave Jordan
Zuzana Urbanek

Dr. Frank J. Pintozzi, Project Coordinator

American Book Company
PO Box 2638
Woodstock, GA 30188-1383
Toll Free: 1 (888) 264-5877   Phone: (770) 928-2834
Fax: (770) 928-7483   Toll Free Fax: 1 (866) 827-3240
Web Site: www.americanbookcompany.com

ACKNOWLEDGEMENTS

The authors would like to gratefully acknowledge the formatting and technical contributions of Marsha Torrens.

This product/publication includes images from CorelDRAW 9 and 11 which are protected by the copyright laws of the United States, Canada, and elsewhere. Used under license.

Copyright© 2006
by American Book Company
PO Box 2638
Woodstock, GA 30188-1318

ALL RIGHTS RESERVED

The text of this publication, or any part thereof, may not be reproduced or transmitted in any form or by any means, electronic or mechanical, including photocopying, recording, storage in an information retrieval system, or otherwise, without the prior written permission of the publisher.

Printed in the United States of America

MCA-II/GRAD Reading Test
Table of Contents

Table of Contents iii

Preface vii
Test-Taking Quick Tips x

Pretest 1
MCA-II/GRAD Test Pretest - Evaluation Chart 26

Chapter 1   Building Vocabulary & Language Skills   27
Context Clues 28
    Figurative Language 30
    Symbolism 32
    Irony 33
    Mood 34
    Satire 35
Types of Resource Materials 35
    Dictionary 35
    Thesaurus 35
    Encyclopedia 36
    Magazines and Journals 36
    Newspapers 37
    Online Search Engines and Keyword Searches 37
Choosing the Right Resource Material 39
Chapter 1 Summary 40
Chapter 1 Review 41
    Web Sites 44

Chapter 2   Main Ideas, Passage Analysis, and Inferences   45
The Main Idea 45
    A Directly Stated Main Idea 46
    An Implied Main Idea 46
Summarizing 49
Paraphrasing 49
Theme In Nonfiction 50
Inferences 53
Chapter 2 Summary 57
Chapter 2 Review 58

Table of Contents

Web Sites 63

Chapter 3 Argument, Audience, Purpose, and Credibility   65

Why We Write 65
Author's Purpose 66
Arguments 68
Voice, Clarity & Style 70
Credentials: Building Trust 71
    Agenda 72
Making Inferences on Bias and Theme 73
Telling Facts and Opinions Apart 76
Audience awareness 77
    Audience Interest 78
    Audience Knowledge 78
    Audience Vocabulary 78
    What the Audience Should Know 79
    Formal and Informal Language 79
Chapter 3 Summary 81
Chapter 3 Review 82
    Web Sites 85

Chapter 4 Literary Elements   87

Theme 88
    Finding The Theme In Fiction 88
Setting 90
Characterization 91
Point Of View 94
    Features of Point of View 94
Tone In Fiction 95
Plot 95
    Subplots 97
Chapter 4 Summary 99
Chapter 4 Review 100
    Web Sites 105

Chapter 5 Consumer, Public, and Workplace Documents   107

Informational Materials 107
Analysis of Workplace Documents 111
    Structure 111
    Format 112
    Graphics 114
    Synthesizing the Information 115
    Headings 115
    The Job Application Process 117
Chapter 5 Summary 120
Chapter 5 Review 121
    Web Sites 123

Practice Test One 125

Practice Test Two 147

Games and Activities 171
Vocabulary (Word Meaning) 171
Reading Comprehension 172
Analysis Of Literature 173

Index 175

MCA-II/GRAD Reading Test
Preface

Passing The MCA-II/GRAD Reading Test will help students who are learning or reviewing standards for the reading sections of the **Minnesota Comprehensive Assessment-II (MCA-II)** exams. The materials in this book are based on the MCA-II /GRAD assessment standards as published by the Minnesota Department of Education.

This book contains several sections:

1) General information about the book itself

2) A pretest

3) An evaluation chart

4) Five chapters that teach the concepts and skills needed for test readiness

5) Two practice tests

Standards are posted at the beginning of each chapter as well as in a chart included in the answer manual.

We welcome comments and suggestions about this book. Please contact the authors at

American Book Company
PO Box 2638
Woodstock, GA 30188-1383

Call Toll Free: (888) 264-5877
Phone: (770) 928-2834
Toll Free Fax: 1 (866) 827-3240

Visit us online at
www.americanbookcompany.com

Preface

About the Project Coordinator:

Dr. Frank J. Pintozzi has taught English and Reading at the high school and college levels for over 27 years. An adjunct professor of Reading and ESL at Kennesaw State University, he has authored eight textbooks on reading, writing strategies and social studies. He holds a doctorate in Education from North Carolina State University Raleigh.

About the Authors:

Michael Kabel was English & Language Arts Director for American Book Company. He received his Masters of Fine Arts in Writing from the University of New Orleans and has worked in media, public relations and publishing for fourteen years. His original fiction has appeared in numerous print and online publications, including *JMWW* and *The Baltimore Review*.

David Jordan holds a Masters in Middle School Education from the University of Georgia. He has taught first and fifth grade students and now divides his time between education writing and working as sports editor for *The Chatsworth Times* in northern Georgia.

Zuzana Urbanek has taught English as a foreign language abroad and in the United States, both to native speakers and as a second language. She holds a Master's degree in English from Arizona State University.

Introduction To The Minnesota Grade 10 Reading Test

The Minnesota Comprehensive Assessment Test in Reading is administered to all Minnesota students enrolled in grade 8 by 2005 – 06 or later. This book offers complete preparation for that test and meets all content standards as described by the Minnesota Department of Education.

In this book, you will prepare for the MCA-II Grad Reading test. First, you will take a pretest to determine your strengths and areas for improvement. In the chapters, you will learn and practice the skills and strategies important to preparing for the test. The last section contains two practice tests that will provide further preparation for the actual test.

Frequently Asked Questions

Will the MCA-II Grad be used to determine promotion for tenth graders?

Successfully completing the MCA-II test is required in order to receive a high school diploma in the state of Minnesota.

What is tested?

The reading test checks your literary comprehension, understanding of literary devices, knowledge of technical documents, and information comprehension skills.

When do I take the Minnesota Reading Test?

Students will take the MCA-II in late April or early May of their 10th grade year.

How much time do I have to take the exam?

There is no time limit.

When will I get the results?

Schools will notify students of their results at an appropriate time.

Where can I find test information online?

The web site of the Minnesota Department of Education is www.education.state.mn.us. It contains test information.

Preface

Test-Taking Quick Tips

1. **Complete the chapters and practice tests in this book.** This text will help you review the skills for the MCA-II/Grad Test in Reading. The book also contains materials for reviewing skill standards established by the Minnesota Department of Education.

2. **Be prepared.** Get a good night's sleep the day before your exam. Eat a well-balanced meal that contains plenty of proteins and carbohydrates.

3. **Arrive early.** Allow yourself at least 15 – 20 minutes to find your room and get settled before the test starts.

4. **Keep your thoughts positive.** Tell yourself you will do well on the exam.

5. **Practice relaxation techniques.** Talk to a close friend or see a counselor if you stress out before the test. They will suggest ways to deal with anxiety. Some other quick anxiety-relieving exercises include:

    1. **Imagine yourself in your most favorite place.** Sit there and relax.
    2. **Do a body scan.** Tense and relax each part of your body starting with your toes and ending with your forehead.
    3. **Use the 3-12-6 method.** Inhale slowly for 3 seconds. Hold your breath for 12 seconds, then exhale slowly for 6 seconds.

6. **Read directions carefully.** If you don't understand them, ask the proctor for further explanation before the exam starts.

7. **Use your best approach for answering the questions.** Some test-takers like to skim the questions and answers before reading the problem or passage. Others prefer to work the problem or read the passage before looking at the answers. Decide which approach works best for you.

8. **Answer each question on the exam.** Unless you are instructed not to, make sure you answer every question. If you are not sure of an answer, take an educated guess. Eliminate choices that are definitely wrong and then choose from the remaining answers.

9. **Use your answer sheet correctly.** Make sure the number on your question matches the number on your answer sheet. If you need to change your answer, erase it completely. Smudges or stray marks may affect the grading of your exams, particularly if the tests are scored by a computer. If your answers are on a computerized grading sheet, make sure the answers are thoroughly dark. The computerized scanner may skip over answers that are too light.

10. **Check your answers.** Review your exam to make sure you have chosen the best responses. Change answers only if you are absolutely sure they are wrong.

MCA-II/GRAD Reading Test
Pretest

The purpose of this pretest is to measure your progress in reading comprehension and critical thinking. This pretest is based on the Minnesota standards for English and Language Arts and adheres to the sample question format provided by the Minnesota Department of Education.

General Directions:

1. Read all directions carefully.

2. Read each question or sample. Then choose the best answer.

3. Choose only one answer for each question. If you change an answer, be sure to erase your original answer completely.

4. After taking the test, you or your instructor should score it using the answer key that accompanies this book. Then determine if you are prepared for the reading comprehension and critical thinking skills tested on the MCA-II/Grad Reading Test.

Read this selection. Then answer the questions that follow it.
Questions 1 – 5 refer to the following passage:

Schedule-Ping™ Installation
Keeping you on schedule!

SYSTEM REQUIREMENTS

*300 MHz or greater processor
*64 MB RAM
*compatible video and audio cards
*4X or higher CD-ROM drive

Introduction

Welcome to the world of Schedule-Ping™, the most complete and user-friendly electronic scheduler available today! You'll be amazed at how this innovative tool helps you keep track of everything in your life. When you use Schedule-Ping™ on a regular basis, you can say goodbye to paper calendars, reminder calls, and all those little notes you leave everywhere for yourself — this software does it all for you!

Installing the Software

To install Schedule-Ping™, insert the CD into your CD-ROM drive. If the launch window does not appear, go to your start menu and choose "Run." Type the letter of your CD-ROM drive, then the word "setup." Follow the on-screen instructions to complete installation.

Once you have completed installation, you will be asked to reboot your computer. The software will be fully installed when the computer is rebooted.

Running the Software

To use Schedule-Ping™, make sure the software is installed (following the directions above) and that your computer has been rebooted. There should be a shortcut icon on your desktop. If the shortcut does not appear, go to your start menu, left click, go to programs and the Schedule-Ping™ folder, and click on the application icon.

Fine Print:

Copyright laws prohibit reproducing the software or the software manual in whole or in part without the permission of the copyright holder.

In the event of a malfunction or other problem arising as a result of faulty manufacturing, the manufacturer will replace the product or issue a refund, at the manufacturer's discretion. However, the manufacturer bears no other responsibility.

In no event will the manufacturer be liable for any financial damage or loss of profits arising out of the use of the software supplied herein.

Due to continued efforts to upgrade and provide improvements, the software specifications may be changed without notice.

For technical support, contact Schedule-Ping™ Customer Service Center at 1-800-555-4545. There is a 90-day warranty period, after which technical support calls are billed at $25/hour. Please have a credit card ready.

1. According to the directions for installation, you will be asked at one point if you want to reboot the computer. What will happen if you choose not to reboot automatically? I.C.3
   A. Nothing; the program is completely installed and ready to use at that point.
   B. You will need to reboot manually for the software to complete installation.
   C. Choose "Run" and type the letter of your CD-ROM drive, then the word setup.
   D. You will have to call customer service to find out how to start the program.

2. The last section of the instructions says that if the software does not work properly, the manufacturer will "replace the product or issue a refund, at the manufacturer's discretion." What does this mean? I.B.G6 I.B.2
   A. It is the software company's choice whether to replace or refund faulty software.
   B. The manufacturer doesn't have to provide a replacement or refund no matter what.
   C. Even if the issue is not their fault, the company will replace or refund the product.
   D. The company will send new software or a refund in a discreet and private manner.

3. Which answer best describes the main idea of the introductory paragraph about the software? I.C.5
   A. When you use this software, you should throw away all calendars and sticky notes.
   B. This software is so amazing that it will keep you on time and in control of your life.
   C. The software is an electronic scheduling replacement for calendars and reminders.
   D. Using this software requires special tools and skills.

4. Under the system requirements for this software, which of the following items needs more clarification for the user to fully understand what is needed? I.C.8
   A. 300 MHz or greater processor
   B. 64 MB RAM
   C. compatible video and audio cards
   D. 4X or higher CD-ROM drive

5. Based on the directions provided, what would you do if you installed the software but did not see an icon on the desktop to open and use the program? I.C.3
   A. Go to your start menu, choose "Run" and type in the name.
   B. Reboot the computer and look for the desktop icon again.
   C. Open the software from the programs list in the Start menu.
   D. Check all of the folders on the desktop for a hidden icon.

Pretest

> The way people get jobs has changed over the years. It's not your parent's job market any more. Read the following article and tables concerning these issues. Then answer questions 6 – 14.

excerpted from "Looking for Work" by Lee Grenon, *PERSPECTIVES*, Autumn 1998 issue:

Over the past two decades, the unemployed have changed their approach to looking for work. Unemployed job seekers make greater use of job advertisements and personal networks, and less use of formal institutions such as public employment agencies and unions. As well, they are increasingly engaged in either a comprehensive job search involving four or more methods over a typical four-week period, or a restricted job search using just one method. These changes have been most pronounced among the long-term unemployed.

Finding a job involves gathering information on jobs and employers, and offering to provide labor to an employer. Employers, employment agencies, friends or relatives, unions, and job advertisements are all potential sources of information. Tapping into these sources can be an active and formal exercise, or a passive and informal one. Despite the trend away from using institutions (Chart), the annual average number of methods used by unemployed job seekers over a typical four-week period has been relatively stable since 1977: between 1.9 and 2.1.

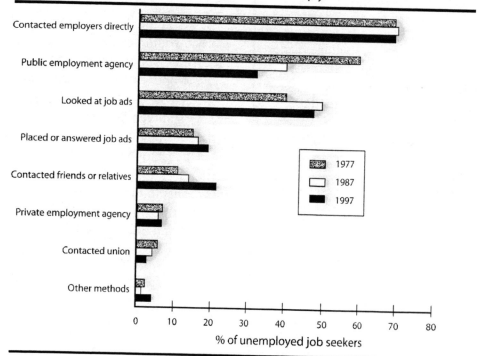

Chart
Use of public employment agencies has declined sharply.

Direct contact most common

As seen in the table on the next page, contacting employers directly has, with few exceptions, remained the most common method of job search over the past 20 years (Table

1), regardless of job seekers' duration of unemployment, age, sex or educational attainment. But the longer the period of unemployment, the less common is direct contact with an employer. Some people may begin to experience job search "burnout" as their list of potential employers shrinks. In 1997, some 71% of job seekers unemployed for 5 to 26 weeks contacted an employer at least once in the four-week reference period, compared with 63% of those out of work for more than a year.

Table 1

Job Search Methods Used by Unemployed Job Seekers

	1977	1982	1987	1992	1997
job seekers	783	1,214	1,120	1,628	1,284
employers directly	537	863	785	1,060	882
public employment agency	470	604	459	582	408
looked at job ads	313	660	550	893	608
placed or answered job ads	114	175	187	294	241
friends or relatives	99	158	173	301	279
private employment agency	40	47	32	55	67
union	31	41	30	44	24
other methods	8	10	15	26	47
	%	%	%	%	%
job seekers	100	100	100	100	100
employers directly	69	71	70	69	69
public employment agency	60	50	41	38	32
looked at job ads	40	46	49	58	47
placed or answered job ads	15	14	17	19	19
friends or relatives	13	13	15	20	22
private employment agency	5	4	3	4	5
union	4	3	3	3	2
other methods	1	1	1	2	4

Over the years, direct contact with employers has become more common among men and younger unemployed job seekers (aged 15 to 24). In contrast, women are now less likely to contact employers directly.

Help wanted

For many years employers have publicized their hiring intentions in newspapers and other publications. This practice now extends to the Internet. The percentage of unemployed job seekers who looked at job advertisements in a four-week period rose until recently. In 1977, 40% did so. The rate increased steadily for more than a decade, reaching 59% in 1993 before dropping back to 47% in 1997. Not surprisingly, this method becomes more common as other methods are exhausted, and as the job seeker approaches "burnout." In fact, the incidence of passive job search (only looking at job notices) has increased steadily over the past 20 years, from 3% in 1977 to 8% in 1997. This growth has been most evident among the long-term unemployed (increasing from 4% to 12%).

6. The final paragraph says that job seekers look at employment ads more often as other methods are "exhausted." In this context, exhausted means:
   A. very tired
   B. used up
   C. finished
   D. damaged

7. According to this report, which of the following is not one of the potential sources of information for people looking for employment?
   A. job ads
   B. flyers
   C. unions
   D. friends

8. What evidence supports the following statement in the third paragraph: "Contacting employers directly has, with few exceptions, remained the most common method of job search over the past 20 years."
   A. from data in the chart
   B. the fourth paragraph
   C. facts about burnout
   D. the data in the table

9. The first paragraph says that changes in job search methods "have been most pronounced among the long-term unemployed." Which of the following words means the same as pronounced in this context?
   A. kept apart
   B. well spoken
   C. very exact
   D. clearly seen

10. Which of the following statements is true, based on the information in the article?
    A. When job seekers get discouraged about contacting employers, they turn to job advertisements.
    B. Women contact employers more often than men.
    C. Looking for new employment always involves multiple steps.
    D. Job seekers look for jobs through friends and relatives more than employment agencies.

11. How many job seekers directly contacted employers in the second year of the study, according to the table?
    A. 71
    B. 100
    C. 863
    D. 1,214

12. Based on the chart, what was the second most utilized method of looking for work in 1997?
    A. public agencies
    B. looking at ads
    C. private agencies
    D. friends and relatives

13. Which of the following statements best describes the main idea?
    A. Unemployed people tap into a wide variety of resources while looking for work, using everything available to them in a well-planned, active, and formal way.
    B. Women who are unemployed use different strategies than men in looking for work, but both of their methods have been found effective in their own way.
    C. Different methods help job seekers look for work, but as each strategy stops working, they experience burnout and have to turn to other resources.
    D. Job seekers have changed their approach to seeking work in the past 20 years, using ads and personal networks more, while using agencies and unions less.

14. If you were gathering information for a report about how people found jobs between 1977 and 1997, what would be your conclusions based on the evidence? Describe the 2 most popular methods in 1977 and 1997. Explain why the methods changed or remained the same.

> Why was the great wall of China built? When was it built? How was it built? For answers to these and other questions, read the following passage. Then answer questions 15 – 21.

Stone by Stone: The Great Wall of China

A very long time ago in the Far East, China's culture was already highly developed and thriving. But with success and prosperity, there often comes division between people. Some struggle to hold on to what they have successfully created, while others struggle even more fiercely to be a part of the success. This is how, in the 7th century BC, the Great Wall of China began.

The wall actually began as a series of simple earthworks — piles of dirt and rock — placed in areas open to invasion. The earliest section of the wall known to be created as a permanent structure was built in the ancient state of Ch'u. Soon, other states saw the building of the wall and began constructing their own earthworks. The wall served as an early warning station for lookouts standing high on platforms. The wall was also easy to defend: arrows and other weapons could be fired from above — down onto any invading forces.

Little is known about the lives of the earliest builders of the wall. It is believed that the wall was built under the direction of military force. But also at that time, there was a cultural expectation that people would work together for the greater good of the community. So the wall may have been a volunteer effort, the result of military intimidation, or a combination of the two.

Each of the different walls built was connected during the Qin dynasty (221– 206 BC). A dynasty is a family of rulers who maintain power over a country for generations. The country of China was for the most part flat, with neither natural boundaries nor natural defensive land forms (mountains, large waterways, etc.). As the country grew, it became more difficult for rulers to defend the land they ruled over. The construction of the wall, which made up for the lack of terrain, grew and improved with each ensuing dynasty.

The materials used in the process of connecting the walls were not mere piles of earth and stones, but brick and granite instead. During the Ming dynasty (1368 –1644), the wall was carefully fortified, or strengthened, with watch towers and designs to both benefit the military and please the eye. These elements of beauty and defense symbolized the splendor and the grandeur of Chinese ingenuity and culture.

15. This article says that, as the Great Wall began, China's culture was highly developed and thriving, but some people struggled to be a part of the success. It also says that the wall was probably built under military force, but at the same time there was a cultural expectation for people to work together for the community. These statements provide examples of what literary element?   I.D.5
    A. satire
    B. symbolism
    C. irony
    D. imagery

16. The main reason that the Great Wall was built was   I.C.5
    A. to unite all the different people living in China.
    B. to symbolize the splendor and grandeur of China.
    C. to protect against invasion.
    D. to separate successful and unsuccessful people.

17. How does the author seems to feel about the Great Wall?   I.C.6 I.C.9
    A. It was built by force, and most Chinese people did not want it.
    B. It is useful as a defense and displays the greatness of China.
    C. It separates and isolates the country from the rest of the world.
    D. It is worth studying scientifically as a very early structure.

18. Which of the following is an opinion, not a fact, about the Great Wall?   I.C.6 I.C.9
    A. "China's culture was already highly developed and thriving."
    B. "The wall actually began as simple earthworks, piles of dirt and rock."
    C. "Other states saw the building of the wall, and they thought that it was good."
    D. "The construction of the wall grew and improved with each succeeding dynasty."

19. The fourth paragraph states, "The construction of the wall grew and improved with each ensuing dynasty." What does ensuing mean?   I.B.G6 I.B.2
    A. following
    B. ruling
    C. stronger
    D. skillful

20. Which of the following is true, based on information in this passage?   I.C.G12
    A. China had to build a wall because it was poor and could not afford to be attacked.
    B. The wall helped form a barrier to outsiders that natural landscape didn't offer.
    C. From the beginning, there was a coordinated effort to join smaller walls being built.
    D. The Chinese did not care what the wall looked like, only that it would be protective.

21. Along with talking about how the Great Wall was built, why does the author include background about China, such as its development, its landscape, and dynasties?   I.D.14 I.D.4
    A. The author is fascinated by the Chinese people more than by the Great Wall itself.
    B. The article would be very short and uninteresting without this added information.
    C. The author was not very organized and changed direction to talk about other things.
    D. The information helps readers understand why, when, and how the wall was built.

A Minnesota company leads the nation on research on the effects of caffeine. Read the following article to find out what they learned. Then answer questions 22 – 30.

excerpted from the *Medical Practitioners' Chronicle*:

The nation's top research labs are conducting studies about the effects of caffeine and phosphates on the health of teenagers. The studies are prompted by increasing reports of nutritional problems among high school and college students. Many nutritionists and physicians blame poor eating habits and increasing stress levels. Caffeine, among other ingredients, was cited as detrimental to the still-growing bodies of young adults.

The Behemoth Chemical Cooperative, headquartered in Mankato, Minnesota, carried out the longest study. It lasted for six months. The study used two groups of volunteer subjects, a study group and a control group. All of the subjects were first-year college freshmen.

The directors gave daily tests in which the subjects responded to various questions about their emotional state and general health. The students answered the questions at the beginning of the day. Before lunch, the directors took blood level readings from the students for caffeine, phosphates, blood sugar, and calcium.

The directors then gave each of the test subjects two twenty-ounce bottles of caffeine-laced soda, which they were to drink at lunch. One hour afterwards, the study directors took more blood level readings and repeated them again three hours later. This last daily session included the directors administering the feelings and emotions test again. The directors put the control group on the same schedule, except this group drank only flavored mineral water without caffeine.

At the end of the study, the directors found high levels of caffeine, phosphates, blood sugar, and calcium in the study group's blood readings. The presence of high levels of calcium meant that the phosphates were stripping the students' bones of calcium. The students' blood sugar readings began to rise and fall more sharply towards the end of the study, suggesting that the increased ingestion of chemicals and sugar was damaging the students' ability to regulate insulin and blood sugar levels.

During the tests, the study group students also began to develop morning headaches, irritability, mood changes, and feelings of ill health. The control group did not report the same experience. The blood level readings for the control group had no large differences for the directors to note.

The federal government has announced plans to run similar studies.

Pretest

22. The first paragraph says, "caffeine is … detrimental to the still-growing bodies of young adults." Which of these words means the same as detrimental?
    A. amusing
    B. destructive
    C. important
    D. harmless

23. What is the purpose of this article?
    A. to warn students about the dangers of drinking caffeine
    B. to educate the public about medical studies about caffeine
    C. to show what happens to students taking part in studies
    D. to report in detail about the study that was conducted

24. Paragraph 7 mentions "increased ingestion of chemicals and sugar." Ingestion means
    A. having something injected.
    B. the beginning of a sequence.
    C. drinking or eating.
    D. feeling bloated after a meal.

25. What did the test subjects do just before drinking two bottles of soda with their lunch?
    A. They had blood level readings taken by study directors.
    B. They answered questions about how they felt that day.
    C. They attended classes, just as they normally would.
    D. They met with the control group to compare results.

26. Why are studies being conducted about caffeine and phosphates?
    A. Students are more active, and scientists want to discover why.
    B. The government is looking to restrict the sales of beverages with caffeine.
    C. Chemical companies want to discover the effects of these ingredients.
    D. Experts want to show that these are unhealthy ingredients for young adults.

27. What evidence in this study links caffeine and phosphates to negative effects on health?
    A. Since phosphates strip calcium from the bones, the students in the test group started feeling weak and in poor health.
    B. Only the test group, which had high levels of caffeine and phosphates in the blood, had symptoms like headaches and irritability.
    C. The students in the test group probably could not sleep due to the caffeine, so that was what caused irritability and mood changes.
    D. Having to drink two whole bottled of soda at lunch made the test group likely to eat less food, and so they became weak and sick.

28. What is something we do *not* know about the subjects in this study?
    A. how long they participated in the study
    B. that they're all first-year college students
    C. how many students were in each group
    D. what they had to do as part of the study

29. What would be a good name for this article?
    A. "Cramming for a test? Caffeine might seem like a good idea but watch out"
    B. "Behemoth Chemical runs longest study yet about caffeine and phosphates"
    C. "Latest study confirms that caffeine and phosphates can harm young adults"
    D. "Should first-year college students be allowed to participate in medical studies?"

30. Which of the following statements best describes the argument in this passage?
    A. The writer uses strong evidence and supporting details.
    B. The writer explores two sides of the issue.
    C. The writer uses the personal testimony of addicted people.
    D. The writer jumps to a conclusion.

End of Section 1. Check your work.

Section 2

> What do you know about the life of an octopus and squid? Yes, they live in the ocean. Learn even more about these shy creatures. Read the following article. Then answer questions 31 – 38.

Shy Shadows: The Octopus and Squid

As I watch, a shadow glides across the ocean floor, where it disappears into rocks which are pockmarked with holes and craters. I have just had a close encounter with the world's first carnivorous predator. No, it is not the shark. My encounter has been with an octopus.

Biologists have identified a class of the Phylum Mollusca (mollusks) as the first active, carnivorous predatory animals. The class is named cephalopoda (Latin for "head-foots"). The class cephalopoda is limited to marine animals and has 600 living species, including species of both octopi and squid.

Luckily for me, the appetites of the two largest mollusks do not usually make them look at humans as dinner. They prefer to dine on fish and crabs, being in the first active, or hunting, family of animals. As predatory animals, the squid and the octopus display unique behaviors and body systems.

Both of these marine animals are shy and choose to stalk their prey quietly. When threatened, the octopus has been observed actually growing pale in color. This behavior provides an illusion of greater size and is a defensive action for the octopus. The octopus can also change colors to blend into the background around it. A different defensive behavior or strategy that the octopus and the squid share with cephalopods is the ability to squirt a cloudy fluid which cuts down on the ability of possible attackers to see them.

These strategies, as well as the octopus' tendency to hide in enclosed spaces (rocks, sunken ships, or reefs), are necessary for survival. Members of the class of cephalopods may have either an external shell, an internal shell, or as in the case of the octopus — no protective shell for the body at all. So the brain, the stomach, the arms, and all the other soft tissues are exposed and vulnerable to injury. The squid has an internal shell which gives it the distinctive streamline shape. The unprotected octopus has a rounded almost undefined shape, undefined, that is except for the body feature that people think of when they think of an octopus or squid.

The arms are the most recognizable features of the octopus and the squid. The octopus' name comes partly from the number of its arms: octo is the Latin word for "eight." Both the octopus and squid use their arms for capturing and eating food. They bring the food to their beak-like jaws to cut it into smaller pieces. The octopus uses its arms also to move along the ocean floor as well as using a flow of water through its body to propel it the along. It can take water into its body and then squirt the water out, propelling itself at a rapid pace. The squid differs only in that it does not use its arms to move. Instead, the squid stays suspended between the ocean floor and surface and moves only by the method of water propulsion. In fact, since the squid has no other means of movement, its body shape and muscles have become well suited to the water propulsion method, making it one of the fastest invertebrate (animal with no backbone) marine animals.

The squid is also one of the largest marine animals. Or rather, the Architecuthis, the giant squid, is the largest invertebrate. There are theories that the old stories of sea monsters were spun by sailors who had seen giant squid. These squid can grow up to 70 feet (21.3 meters), and they have the largest eyes of any animal on Earth. Oddly enough, no one has ever seen a living giant squid in its natural environment. Either it hides well or the stories of sea monsters have a hold on even the most adventurous sea explorers. However, they do wash up dead on beaches or are caught in fishing nets. Recently a 28-foot squid was caught by a trawler off the coast of the Falkland Islands.

Keeping an eye out for the marine creatures which do not turn down humans for lunch, I slowly head towards the rocks where the shadow has hidden. My time is running short, along with the air in my oxygen tanks. I must go back to my own world, away from the hushed, dim world of the cephalopods. Before I go, I catch a last glimpse of the creature, turned a reddish color to match the rocks behind it. The fragile octopus has every reason to be shy. It has no shell to retreat into.

31. Why does the author call octopi and squid "shy" creatures?
  A. When confronted, they turn different colors, just like people blush.
  B. If threatened, they retreat or act defensively.
  C. When other creatures look at them, they always close their big eyes.
  D. They are always alone.

32. In the introduction, the author says, "a shadow glides across the ocean floor." Why does the author use this metaphor rather than simply naming the creature right away?
  A. The author wants the creature to remain a mystery until he reveals it at the very end.
  B. The figurative language paints a picture and makes the reader want to find out more.
  C. The author wants to create a dark atmosphere to talk about very scary sea creatures.
  D. The word "shadow" is another name for octopi that live near the bottom of the sea.

33. How does the author feel about squid and octopi?
    A. The author really does not reveal any feeling about the creatures.
    B. The author seems to be deathly afraid of these "behemoths."
    C. The author sees these creatures as interesting for scientific study but remains unemotional.
    D. The author almost gives them personalities and is fascinated by them.

34. Which of these answers is an example of evidence that the author provides to support that octopi and squid have "unique behaviors and body systems"?
    A. "Luckily for me, the appetites of the two largest mollusks do not usually make them look at humans as dinner."
    B. "The arms are the most recognizable features of the octopus and the squid."
    C. "There are theories that the old stories of sea monsters were spun by sailors who had seen giant squid."
    D. "The fragile octopus has every reason to be shy. It has no shell to retreat into."

35. Which animal can squirt a cloudy substance to keep its attackers from seeing it clearly?
    A. octopus
    B. squid
    C. both
    D. neither

36. Which animal can travel very fast through the water?
    A. octopus
    B. squid
    C. both
    D. neither

37. The article points out that no one has seen a living giant squid in its natural environment. How does the author know they exist?
    A. He knows from old stories of sea monsters told by sailors who'd seen giant squid.
    B. Sometimes they wash up dead on beaches or get caught in fishing nets.
    C. He sees a giant squid descend to the ocean floor as he scuba dives in the ocean.
    D. There is one on display at a museum where the author did some of his research.

38. In which of the following lines does the author reveal how many arms an octopus has?
    A. "Both the octopus and squid use their arms for capturing and eating food."
    B. "The arms are the most recognizable features of the octopus and the squid."
    C. "The octopus' name comes partly from the number of its arms: octo is the Latin word for 'eight.'"
    D. "The octopus uses its arms also to move along the ocean floor as well as using a flow of water through its body to propel it along."

> Only rarely do you hear of a man who truly inspired people by his life. Booker T. Washington is one of those people. Read about him in the following article. Then answer questions 39 – 46

Booker T. Washington: Struggle and Triumph

One of the most inspiring "rags to riches" stories is absolutely true. A child was born a slave in 1856. As a slave, he wasn't allowed to go to school. After the Civil War and the Emancipation Proclamation, his only dream was to get an education. In 1872, Booker Taliafero Washington made his way to the Hampton Institute in Virginia. Though he had neither money nor references, he had a clear desire to learn and to help others learn. Luckily, he had come to the right place: the Institute was dedicated to training black teachers.

Tuskeegee Institute

Booker T. Washington was a successful student and was quickly employed as a teacher. When a black college, the Tuskeegee Institute, was built in Alabama, Booker T. Washington became its first principal and a champion of the school's purpose in training black men to be self-sufficient.

Washington proved to be a charismatic leader. He was a handsome man who kept himself physically fit and spoke well to individuals and large groups alike. He was able to translate his vision for the college into words, and to raise a great deal of money for the Institute's growth. One of the people he enlisted to help was steel-making tycoon Andrew Carnegie, a man who had also risen from poverty to become successful. Carnegie visited the college and helped to fund many of Washington's ideas for expanding the Institute's work. Washington also recruited leaders in other fields. George Washington Carver, an inventor, botanist and one of the first nationally-respected black scientists, was among the Institute's earliest guests.

Booker T. Washington

Washington was also the first black man to dine at the White House, at the invitation of President Theodore Roosevelt. The two men shared many views, including the need for healthy physical activity, the value of books and knowledge, and the redeeming power of morality.

In his time, Booker T. Washington was a great leader for his people; encouraging them to learn skills to become economically equal with the white race. He did, however, believe in the system or philosophy of the races being kept separate, earning the nickname "The Great Accommodator" from his critics. Other black leaders grew impatient with the slow pace of change and urged stronger measures to gain equality.

Washington was on a speaking tour in New York City when he fell ill. He asked to be taken home to the South to be buried. He died at home in Tuskeegee, where he is buried near the Institute's chapel.

39. Based on the passage, which answer best describes the man, Booker T. Washington? I.D.6 I.D.10
    A. He was a shy and unassuming man who wrote convincingly for social change.
    B. Born into a wealthy family, he felt fortunate and wanted to give something back.
    C. Since his efforts in the South were ineffective, he left to work with politicians.
    D. He vigorously pursued a better life for his people.

40. What three genres of literature would this selection best fit? I.D.7
    A. memoir, narrative, autobiography
    B. diary, history, travel literature
    C. historical fiction, drama, saga
    D. history, biography, narrative

41. What is the main conflict in this passage, and how is it resolved? I.D.6 I.D.10
    A. Washington had trouble running the college, so he ended up leaving the South.
    B. Washington had always wanted to visit the White House, and President Roosevelt invited him there.
    C. Washington's poverty made him long for education and a better life, which he achieved and secured for many others.
    D. Washington wanted to be a good student, and he did so well that he became a teacher.

42. The mood and tone of the piece are best described as I.D.6 & I.D.10
    A. happy.
    B. sad.
    C. hostile.
    D. objective.

43. The first sentence calls this a "rags to riches" story. What is the literal meaning of this expression? I.D.5
    A. Going from very poor circumstances to a valuable condition.
    B. Something that was run-down being renovated with lots of money.
    C. Something that was worn to shreds can become like new again.
    D. Comparing the way poor people dress to the attire of rich people.

44. The theme of the passage is best expressed in which of the following? I.D.14
    A. Washington shows great courage in meeting the president and other important people.
    B. Even someone as unfortunate as a slave can achieve great things in life.
    C. Washington had no money, but he managed to talk other people into giving it to him.
    D. People who were once poor, like Washington and Carnegie, can help each other.

45. How did Booker T. Washington become free from slavery? I.C.G12 I.C.7
    A. He escaped and applied for free citizenship.
    B. Slavery was abolished after the Civil War.
    C. Becoming a student automatically freed him.
    D. He earned enough money to buy freedom.

46. On your own paper, write a response in which you discuss the development of the author's perspective about Book T. Washington. Use specific examples from the text to support your ideas. I.C.6 I.C.9 I.D.6 I.D.1

> Did you ever wonder what kind of movies your grandparents and great grandparents went to when they were kids? Read about the old movies and how they've changed. Then answer questions 47–52.

Gangsters & Cowboys: American Films Reflect Their Times

By the height of the Great Depression in 1933, more Americans were flocking to movies than ever before, eager to escape the crushing poverty and pervading sense of hopelessness that massive unemployment and growing debt brought on a seemingly daily basis. Because tickets cost only pennies and offered hours of diversion, they existed for millions of Americans as a cheap way to fill the hours of the day. Moreover, movies had yet to escape the public perception as a "working class" medium — fit not for the cultured, educated elite but for the lower, blue collar citizen and people of little, if any, secondary or high school education. Hollywood responded by making films that both appealed to peoples' yearnings for release from economic want and attacked the government and social systems many people held at least partly responsible for the Depression.

A great example of the films that both exploited and glamorized wealth were the hugely popular "Thin Man" series, starring William Powell and Myrna Loy. The two played Nick and Nora Charles, rich socialites who solved murders among New York's elite society. Set in luxurious wardrooms and expensive nightclubs, the murders always involved the idle rich and were motivated by petty emotions of greed and vanity. Powell's Nick, a former working class gumshoe, uses a common sense logic and "from the streets" attitude to see through the pretensions and snobbery of the socialites. In every example, his unpretentious common sense won out over the idle rich's treachery, standing up for justice regardless of economic station.

James Cagney

Similarly, the gangster films that first became prominent during this period showed audiences the modern-day outlaws of that era, individuals who defied the systems responsible for the Depression and made their own way in an uncaring environment. Films such as *The Public Enemy* and *Angels with Dirty Faces* showed working class men and women clawing their way into riches despite the police and courts who stood in their way. Such social institutions were, to many people, symbolic of the same economic and social systems responsible for the Depression. That the criminals often died violently was almost beside the point to audiences who devoured these films. The gangster heroes won out, even in death.

By the 1970s, anxiety regarding the nation's direction and sense of identity became reflected in popular entertainment, as a new generation of writers and directors — products of the social conscience of the previous decade — exerted their influence on the entertainment mainstream. In the aftermath of the Vietnam War and the growing Watergate scandal, American film audiences entered into a new romance with the Western genre, albeit a new strain that served as social allegory. In these "New Westerns," the heroes were frequently lawbreakers and outcasts who rejected social convention and morality in order to make a life for themselves on the American Frontier. Films such as *Butch Cassidy and the Sundance Kid* and *Jeremiah Johnson* (both starring Robert Redford) resonated deeply with audiences worried about confining government and technological structures; *The Outlaw Josie Wales*, about a Confederate soldier (Clint Eastwood) looking to lead a productive life in the West, spoke to the post-Vietnam yearning for individual redemption. Similarly, the brutal violence of vigilante films such as *Dirty Harry* (also with Eastwood) and *Taxi Driver* enthralled audiences beleaguered by skyrocketing crime rates and a weakened justice system. In both types of films, the recurring theme is "don't trust governments or economies; you're on your own."

The theme of the individual against crushing social pressure continued throughout the decade, spiraling into urban political drama (*Serpico*), science fiction (*The Omega Man*), and even romance (*Love Story*). While by the beginnings of the 80s the pessimism had thawed somewhat, the 70s spirit of individualism influenced the next generation of filmmakers who came of age in the 1990s. Directors such as P.T. Anderson, Steven Soderbergh and The Wachowski brothers adapted the outlaw filmmaker creed, crafting anti-establishment works such as *Magnolia*, *Traffic* and *The Matrix* trilogy.

47. What is the main point that the author makes about films in this article?
   A. Film studios only make films that will earn money.
   B. Movies reflect the fears and desires of society at any given time.
   C. Only the movies with big stars are popular with audiences.
   D. People can learn a great deal of history from the movies.

48. How does the author build an argument in this passage?
   A. by using an emotional appeal to influence readers
   B. by including vivid description to illustrate main points
   C. by using clear organization and detailed supporting facts
   D. by offering contrasting views to point out the best choice

49. The author says that, in the 1970s, audiences were "stressed by skyrocketing crime and a destabilized justice system." Destabilized means the system was
   A. helpless.
   B. uneven.
   C. weakened.
   D. excessive.

50. According to the passage, what kind of movie hero became popular in the 1970s?
   A. lawbreakers and outcasts
   B. cops and robbers
   C. politicians and newsmakers
   D. cowboys and Indians

51. What is the main reason the author gives for booming movie attendance during the Depression?
   A. So many great films were being made that people wanted to see them.
   B. The blue-collar class had no other entertainment available to attend.
   C. Many new film stars were rising and capturing people's attention.
   D. Movies offered an escape from day-to-day poverty and unemployment.

52. According to this passage, the writers and directors who became important in the 1970s were influenced by what events?
   A. the hardship of the Great Depression years
   B. the Vietnam War and the Watergate scandal
   C. government control and new technology
   D. gangster violence and vigilantes acting out

Does your family still peel their own potatoes, or is everything "instant" at your house? Read the following article and see what you're missing. Then answer questions 53 – 58.

"To Peel Potatoes"
excerpted from *Peace Corps: The Great Adventure,* by John P. Deever

"Life's too short to peel potatoes," a woman in the supermarket announced, putting a box of instant mashed potatoes in her cart. When I overheard her, I nearly exploded.

Having recently returned from my Peace Corps stint in the Ukraine, I tend to get defensive about the potato in all its forms: sliced, scalloped, diced, chopped, grated, or julienned; then boiled, browned, french-fried, slow-fried, mashed, baked, or twice-baked —with a dollop of butter or sour cream, yes, thank you.

A large portion of my time in the Ukraine was spent preparing what was, in the winter, nearly the only vegetable available. Minutes and hours added up to days spent handling potatoes. I sized up the biggest, healthiest spuds in the market and bought buckets full, then hauled them home over icy sidewalks.

Winter evenings, when it got dark at four p.m., I scrubbed my potatoes thoroughly under the icy tap — we had no hot water — until my hands were numb. Though I like the rough, sour peel and prefer potatoes skin-on, Chernobyl radiation lingered in the local soil, so we were advised to strip off the skins. I peeled and peeled, pulling the dull knife toward my thumb as Svetlana Adamovna had taught me, and brown-flecked stripe after stripe dropped off to reveal a golden tuber beneath. Finally I sliced them with a "plop" into boiling water or a hot frying pan. My potatoes, my kartopli, sizzled and cooked through, warming up my tiny kitchen in the dormitory until the windows clouded over with steam.

Very often my Ukrainian friends and I peeled and cooked potatoes together, either in my kitchen or in Tanya's or Misha's or Luda's, all the while laughing and talking and learning from each other. Preparing potatoes became for me a happy prelude to food and, when shared with others, an interactive ritual giving wider scope and breadth to my life.

But how could I explain that feeling to the woman in a grocery store in the United States? I wanted to say, "On the contrary, life's too short for instant anything."

Back home, I'm pressed by all the "instant" things to do. In the Ukraine, accomplishing two simple objectives in one day — like successfully phoning Kiev from the post office and finding a store with milk — satisfied me pretty well. I taught my classes, worked on other projects, and tried to stay happy and healthy along the way.

Now it takes an hour of fast driving to get to work, as opposed to twelve minutes of leisurely walking. I spend hours fiddling with my computer to send "instant" e-mail. Talking to three people at once during a phone call is efficient — not an accident of Soviet technology. With so much time-saving, I ought to have hours and hours to peel potatoes. Somehow I don't.

What I wish I'd said to the woman in the supermarket is this: "Life's too short to be shortened by speeding it up."

But I wasn't able to formulate that thought so quickly. Instead, I went to the frozen food section and stared at the microwave dinners for a while, eventually coming to the sad, heavy realization that the Szechuan chicken looked delicious — even if it didn't come with potatoes.

53. What does the author seem to miss most about the Ukraine, even with all of its difficulties?
    A. the grocery markets
    B. the slower pace
    C. the friends he made
    D. peeling potatoes

54. As the title of the passage states, it is about potatoes on the surface, but what is the underlying theme?
    A. Life is too short to peel potatoes.
    B. There's never enough time to shop.
    C. Take time to enjoy even small tasks.
    D. Potatoes can be prepared in many ways.

55. What point of view is used to tell this story, and why was it chosen?
    A. It is presented in second person, as if the reader were actually there.
    B. It is related in third person, so we can get to know all the characters equally.
    C. It is told in first person, as the author is relating his own personal experience.
    D. It is written in third person, which allows the facts to be presented objectively.

56. The author writes, "Preparing potatoes became for me a happy prelude to food." A prelude is
    A. an event leading up to something else.
    B. a substitute for anything that is not in season.
    C. something served alongside a main dish.
    D. a difficult task rewarded by something pleasant.

57. The tone that the author uses to talk about this topic is
    A. humorous.
    B. disinterested.
    C. dreadful.
    D. passionate.

58. The author goes into detail about potatoes and their preparation. On your own paper, analyze why and how this is done, mentioning specific imagery used.

End of Section 2. Check your work.

Section 3

Read this account of a day in the life of a soldier in training. Read how making a mistake can teach someone a lesson for life. Then answer questions 59 – 65.

A Soldier's Lesson
by Dick Regis

It was in the sixth week of basic training, and I was getting quite tired during our mock battlefield training exercise. After a six-mile march in full gear, we had to dig our foxholes (underground defensive positions) and settle in for the night. The afternoon sun was beating hard, and we had to dig with a very small shovel known as an entrenching tool.

As we dug, my platoon buddy, Private Nixen, turned to me and said, "Hey! We need branches to camouflage our foxhole. Why don't you go get them?"

"Sure," I said. "As long as you can watch my M-16 rifle. You know how the enemy unit is always trying to outwit us. They may try to steal it from me as I gather wood."

"No problem," Nixen said.

Private Bryant, who was within earshot of our foxhole, shouted, "Be careful out there — you never know when the enemy is going to strike."

So, off I went to gather branches loaded with leaves. This task took longer than I expected. It took me almost thirty minutes to get back to camp and make my way over to our foxhole. I was glad to be back in one piece and dropped the heavy load. However, what happened next made me wish that I had stayed away.

My Drill Sergeant, Oglethorpe, was standing next to Nixen and had a look on his face that would melt mountains. He thundered, "Private, do you know where your M-16 is? Do you?"

My eyes darted frantically, but I didn't see my weapon anywhere.

"Drill Sergeant, I don't know where it is," I stammered. I looked over at my buddy, but Nixen remained silent.

"Well, the enemy unit returned it to me. They said they had found the rifle unattended, and so they took it. Do you have any idea how embarrassing that is to me?" Oglethorpe thundered.

"It is my fault," I said. "I know, ultimately, the rifle is my responsibility." Once I said this, my drill sergeant shared some unsavory words with me, ordered me to hold my rifle in front of me for thirty minutes, and stomped off. Each minute of standing like that seems like an excruciating week, and my arms were sore for a week after this exercise.

The next day, Friday, the platoon returned to the barracks. Everyone was allowed to go off post on leave Friday and Saturday and return Sunday. Unfortunately, I was not permitted to leave the barracks for the weekend. Because of the loss of my rifle, I had to stay behind and clean the latrines of my barracks. However, I suffered through this and just considered it one of life's memorable experiences.

I still remember Nixen's look of relief and happiness when he left for his two day leave.

59. What seems to be the main purpose of the author in writing this story?
    A. to share a personal experience
    B. to introduce a character
    C. to create a mood
    D. to entertain

60. When the narrator returns with an armload of branches, he says that the sergeant looks at him with a face that would "melt mountains." What does that most likely mean?
    A. The sergeant's face is hard as a rock, resembling a mountain range.
    B. The sergeant's face is red and hot enough with anger to melt stone.
    C. The sergeant was so absorbed in paperwork, his eyes seem melted.
    D. The sergeant is so thirsty, like his men, melting anything would help.

61. What is the first thing that the narrator does when he gets back to the foxhole?
    A. He searches around for his rifle.
    B. He apologizes for taking so long.
    C. He sees the sergeant looking at him.
    D. He drops the branches he had gathered.

62. What kind of a person does the sergeant seem to be?
    A. gentle and easily embarrassed
    B. frightened and apprehensive
    C. curious and always meddling
    D. intimidating and demanding

63. The author says that, before he was punished, the drill sergeant shared some <u>unsavory</u> words with him. In this context, <u>unsavory</u> most closely means
    A. disciplinary.
    B. bland.
    C. unpleasant.
    D. brave.

64. Why is the narrator not more upset about having to stay on base and clean latrines?
    A. He really did not want to go on off post and spend all his money.
    B. He realizes the punishment fits the way he ignored his responsibility.
    C. He knows that if he tried to leave, the punishment would be worse.
    D. He is relieved that no one was hurt as a result of him losing his rifle.

65. What is the climax of the action in this story, and what elements does the author use to make it effective? Answer on your own sheet of paper.

This poem from another culture gives insight into what these people value in life. Read the poem. Then answer questions 66 – 69.

The Singer's Art
an Aztec Poem

1 I polish the jade to brilliance
   I arrange the black-green feathers
   I ponder the roots of the song
   I order in rank the yellow feathers,
5   So that a beautiful song I sing.
   I strike the jade continuously
   To break out light from the flower's blossoming,
        Only to honor the Lord of the Close and the Near.

   The yellow plumes of the *troupial**,
10 The blue-green of the *trogon**,
   The crimson of the rosy spoonbill.
   I freshly arrange.
        My noble song, sounding golden tones,
        My song I will sing.
15 A golden finch my song is proclaimed.
   I sing it in the place of the raining flowers,
        Before the face of the Lord of the Close and the Near.

*The *troupial* and the *trogon* are names of tropical birds.

66. What can you interpret about the poet's emphasis in the poem above?

   A. The poet focuses on conflict with nature.
   B. The poet wants to rearrange nature's design.
   C. The poet describes the harmony of a bird's feathers and voice.
   D. The poet is concerned with preserving birds for future generations.

67. What do lines 1 – 5 highlight about the author?

   A. The poet sees nature posing a beautiful pattern.
   B. The poet finds comfort from planting trees with strong roots.
   C. The poet believes in embracing nature as a beautiful, chaotic force.
   D. The poet enjoys making jade jewelry for other Aztecs.

68. What does the figurative language in lines 12 – 14 emphasize?   I.D.4

   A. Having nicely placed feathers affects its creativity.
   B. It gives the bird pride in itself and its song.
   C. It makes the other birds jealous.
   D. The Lord of the Close and the Near appreciates beautiful birds.

69. Which line from the poem contains alliteration?   I.D.5

   A. Line 1
   B. Line 9
   C. Line 12
   D. Line 14

Read the following poem and see how the life of a hard working blacksmith is seen through the eyes of the poet. Then answer questions 70 – 72.

"The Village Blacksmith" by Henry Wadsworth Longfellow

1 Under a spreading chestnut-tree
   The village smithy stands;
   The smith, a mighty man is he,
   With large and sinewy hands;
   And the muscles of his brawny arms
   Are strong as iron bands.

7 His hair is crisp, and black, and long,
   His face is like the tan;
   His brow is wet with honest sweat,
   He earns whate'er he can,
   And looks the whole world in the face,
   For he owes not any man.

13 Week in, week out, from morn till night,
   You can hear his bellows blow;
   You can hear him swing his heavy sledge,
   With measured beat and slow,
   Like a sexton ringing the village bell,
   When the evening sun is low.

19 And children coming home from school
   Look in at the open door;
   They love to see the flaming forge,
   And hear the bellows roar,
   And catch the burning sparks that fly
   Like chaff from a threshing-floor.

25 He goes on Sunday to the church,
   And sits among his boys;
   He hears the parson pray and preach,
   He hears his daughter's voice,
   Singing in the village choir,
   And it makes his heart rejoice.

31 It sounds to him like her mother's voice,
   Singing in Paradise!
   He needs must think of her once more,
   How in the grave she lies;
   And with his haul, rough hand he wipes
   A tear out of his eyes.

37 Toiling,—rejoicing,—sorrowing,
   Onward through life he goes;
   Each morning sees some task begin,
   Each evening sees it close
   Something attempted, something done,
   Has earned a night's repose.

43 Thanks, thanks to thee, my worthy friend,
   For the lesson thou hast taught!
   Thus at the flaming forge of life
   Our fortunes must be wrought;
   Thus on its sounding anvil shaped
   Each burning deed and thought.

70. Which of the following dictionary definitions best defines the word <u>sinewy</u> as it is used in the poem? I.B.2
    A. possessing excess weight
    B. stringy and tough
    C. lean and muscular
    D. (of meat) full of sinews; especially impossible to chew

71. What do lines 7 – 12 highlight? I.D.14
    A. He is wealthy.
    B. He is a hard worker.
    C. He is old.
    D. He doesn't like children.

72. Which of the following best explains the internal conflict of the blacksmith? I.D.6, I.D.10
    A. He is sad to see his children grow up.
    B. He is tired of working but can't afford to retire.
    C. He feels guilty about once being dishonest.
    D. He misses his wife but must keep on going.

End of Section 3. Check your work.

MCA-II/GRAD Test Pretest

Evaluation Chart

Directions: On the following chart, circle the question numbers that you answered incorrectly and evaluate the results. Then turn to the appropriate topics (organized by chapters), read the explanations, and complete the exercises. Review other chapters as necessary. Finally, complete the **two practice tests** to further prepare yourself for the Minnesota MCA-II/GRAD tests.

Chapter	Content Standards	Questions
Chapter 1: Building Vocabulary & Language Skills	I.D.5, I.B.2, I.B.G6	2, 6, 9, 15, 19, 22, 24, 32, 33, 43, 49, 55, 56, 57, 58, 60, 63, 68, 69, 70
Chapter 2: Main Idea, Passage Analysis, and Inferences	I.C.7, I.C.5, I.C.G11, I.C.G12	3, 7, 8, 10, 11, 12, 13, 16, 20, 25, 26, 27, 29, 35, 36, 37, 38, 45, 47, 50, 51, 52, 53, 54, 61, 64
Chapter 3: Argument, Audience, Purpose, and Credibility	I.C.6: & I.C.9, I.C.8, I.D.4, I.D.7:	4, 14, 17, 18, 23, 28, 30, 34, 46, 48, 55, 59, 66
Chapter 4: Literary Elements	I.D.4, I.D.6, I.D.10, I.D.14	21, 31, 39, 40, 41, 42, 44, 54, 62, 65, 67, 71, 72
Chapter 5: Consumer, Public, and Workplace Documents	I.C.3	1, 2, 3, 4, 5

Chapter 1
Building Vocabulary & Language Skills

This chapter covers the following content standards:

1.B.2	Students will determine the meaning of unfamiliar words and metaphors by using dictionaries, context clues and reference books.
1.B.G6	Students will determine the meaning of unfamiliar words by using context clues.
1.D.5	Students will analyze, interpret and evaluate the use of figurative language and imagery in fiction and nonfiction selections, including symbolism, tone, irony and satire.

Saying someone learns to read while still a child is somewhat misleading. The skill of reading develops as we grow older, so that reading a work at age ten has a different effect on us than when we reread it at age 15. Reading involves the reader mentally, and as we understand a book more, our enjoyment grows deeper. As a result, the books that fascinate a child bore an adult, and vice versa. But the act of reading to understand stays the same, despite differences in age and skill.

No matter how prestigious or simple a book or reading passage is, or the age at which we read, a work's success or failure often depends largely on its mastery of **language**. Language is the words an author uses to convey his ideas, memories and emotions. It is the basic workings of any story, the "moving parts" that allow it to function. Through language, a writer creates **imagery:** the use of vivid language to express ideas, emotions, or objects.

How well you understand language and its many functions largely determines how much you will enjoy a novel, short story, magazine article, or play. If you have ever found yourself confused by the language of Shakespeare, or amused by a Harry Potter novel from J.K. Rowling, you have seen the difference that your command of language can make in your final appreciation of fiction and nonfiction.

Building Vocabulary & Language Skills

The major concepts covered in this chapter include:

context clues	understanding an unfamiliar word by using the words around it
figurative language	figures of speech; words and phrases not meant to be taken literally
symbolism	using objects, people or events to suggest larger meanings
mood and tone	the "soul" of a work and the writer's attitude towards his subject
irony	differences between what is expected and what actually happens
satire	storytelling in which human faults are attacked using wit, irony, and sarcasm

We will also discuss how to use a **dictionary** and other reference sources.

When going through a text, remember that good reading involves allowing your mind to come to its own decisions and trusting your instincts. This book will help improve those instincts, so that your understanding of literature and recreational materials becomes clear and sharp. For now, remember to read "with your mind," asking questions and making observations about what you absorb.

CONTEXT CLUES

Using **context clues** to find the meanings of words involves looking at the way words are used in combination with other words in their setting. By looking at and analyzing the phrases and "signal words" that come before or after a particular word, you can often figure out its meaning. The idea or message of the whole text becomes clearer once we understand the **context**, or circumstances, in which a word occurs.

Look at the words and phrases around an unknown word. Think about the meaning of these words or the idea of the whole sentence. Then match the meaning of the unknown word to the meaning of the known text.

In the following statements, choose the word which best reflects the meaning of the bolded word.

1. Being a realist, Tom was highly **skeptical** of the UFO story.
   A. simple
   B. trustful
   C. impressed
   D. doubtful

2. Exercise that increases your heart and breathing rates for a sustained period of time is called **aerobic**.
   A. of long duration
   B. improving strength
   C. requiring oxygen
   D. strenuous

Chapter 1

In statement 1, the clause "being a realist" suggests that Tom isn't likely to believe fantastic stories. As such, we can imagine he won't automatically trust a story about UFOs. Choice D, "doubtful," best explains this relationship. For statement 2, you may use a **definition clue**. The signal word "is" indicates that the definition of aerobic is described in the first part of the sentence. Since heart and breathing rates are *increased* and all organisms need oxygen to live, aerobic must mean C, "requiring oxygen."

Context Clues	Signal Words
Comparison	*also, like, resembling, too, both, than*
	Look for clues that indicate an unfamiliar word is similar to a familiar word or phrase.
	Example: The accident felled the utility pole like a tree for timber.
Contrast	*but, however, while, instead of, yet, unlike*
	Look for clues that indicate an unfamiliar word is opposite in meaning to a familiar word or phrase.
	Example: Stephanie is usually in a state of *composure* while her sister is mostly boisterous.
Definition or Restatement	*is, or, that is, in other words, which*
	Look for words that define the term or restate it in other words.
	Example: The principle's idea is to *circuit* — or move around — the campus weekly to make sure everything is okay.
Example	*for example, for instance, such as*
	Look for examples used in context that reveal the meaning of an unfamiliar word.
	Example: People use all sorts of *vehicles* such as cars, bicycles, rickshaws, airplanes, boats, and motorcycles.

Practice 1: Using Context Clues

Above each bolded word, write its meaning. Use context clues to help you.

1. Those who cannot afford **bail** cannot be freed on pre-trial release.

2. Hank said the ocean was very **tranquil**; I also thought the ocean was peaceful.

3. Sometimes strong **herbicides** are needed to eliminate weeds from the garden.

4. Residues such as ammonia even show up in grain sprayed with **pesticides**.

5. Since Brian disliked working for others, he decided to become an **entrepreneur**.

6. This word is **ambiguous**; it can have two meanings.

7. Jennifer wanted no **remuneration** in money or gifts; her reward was saving the pet.

8. He was a **fastidious** dresser, always very neat and particular about what he wore.

9. After a **cursory** examination of only a minute or two, the doctor said he did not believe there was anything seriously wrong with the child.

10. Smoking too much is likely to have a **pernicious** effect on one's health.

Some authors and philosophers have argued that everything depends on *context*, the circumstances in which an event occurs. Language is much the same way — the meaning of a sentence, phrase or image is largely dependent upon the circumstances in which it was written. Sometimes, language becomes so *abstract* (apart from concrete experiences) that we can say it's not meant to be taken literally. Such use of words is known as **figurative language**.

FIGURATIVE LANGUAGE

We use **figurative language** constantly in our daily lives, to describe an event or experience we've had, or to convey a feeling or idea with extra emotional impact. Figurative language is used extensively in fiction and poetry, where the author relies on vivid or evocative language to communicate powerful ideas.

Writers work in patterns of words to create their images. They may repeat sounds or words, compare two different things, or create a likeness between words and sounds.

Personification is the device of giving human qualities to something not human. For example, "the stars stare down on us" implies that the inanimate stars watch us as other people might. Other examples include:

"The stop sign jumped out at me."

"Summer waited just around the corner."

"The moon winked at the young lovers."

A **simile** is a comparison of one thing to another, using the words *like* or *as*. Because they make direct connections between two things, similes are said to be *explicit* comparisons:

"The cold wind howled like a starving wolf."

"Her smile was as faint as a fat lady at a fireman's ball" – Raymond Chandler.

"Her skin was as soft as a rose."

Metaphor is the comparison of two objects without using the words *like* or *as*; it is an *implicit* comparison. For example, in Shakespeare's "As You Like It," a character observes

"All the world's a stage

And all the men and women merely players.

They have their exits and their entrances."

Shakespeare connects life itself with a theatrical production, with birth and death akin to the comings and goings of actors in a drama. The speech goes on to list the stages of a man's life as different parts to be played.

An **extended metaphor** continues over several sentences or passages, sometimes over an entire work. George Orwell's *Animal Farm* is an elaborate re-imagining of the 1917 Bolshevik Revolution in Russia. The entire text is meant to represent the rise and ethical decline of the communists' control of Russian society. Works that use metaphor over their entire narrative belong in a special category called **allegory**.

An **allusion** is a reference to a specific place, a historical event, a famous literary figure, real or fictitious, or a work of art. Allusions can be drawn from history, geography, science, math, religion, or literature. In order to identify an allusion in a piece of literature, the reader must have prior knowledge of the reference in question. In the film *The Matrix*, for example, Morpheus alludes to Lewis Carroll's book "Alice in Wonderland" when he tells Neo he can either stay in wonderland or see how far down "the rabbit hole" goes. If you were not familiar with Lewis Carroll's *Alice in Wonderland*, you might not identify the allusion.

You probably use allusions all the time, as modern slang is filled with references to vivid images or famous people, quotations, events or locations. Some examples include:

"more money than Bill Gates."

"a complete Pearl Harbor"

"go postal"

"uglier than a homemade sandwich."

Practice 2: Figurative Language

Read the following sentences. Then identify the figurative language used in each.

1. "I slept like a baby: I woke up every two hours screaming."
   A. simile B. metaphor C. satire D. allusion

2. He was the Tiger Woods of the chess club.
   A. simile B. alliteration C. allusion D. symbolism

3. "I had rather be a canker in his hedge than a rose in his grace." – from *Much Ado About Nothing*
   A. simile B. metaphor C. onomatopoeia D. irony

4. He's pretty much the Quentin Tarantino of his generation.
   A. simile B. allusion C. symbolism D. alliteration

5. The car told me it was running out of gas.
   A. alliteration B. symbolism C. simile D. personification

6. The wind whistled through the cracks in the old cabin.
   A. simile B. slogan C. allusion D. personification

Building Vocabulary & Language Skills

SYMBOLISM

We use **symbols** such as signs, posters, and logos in everyday life, as shorthand for ideas or concepts. Symbols, over time and widespread use, have become universal — recognizable to everyone, regardless of heritage or language. For example, a skull and crossbones represents "poison," while a red hexagon means "stop."

Symbols appear in literature, as well, and while there are *universal symbols* that appear time and again, authors will create their own symbolism to reinforce and deepen their intended themes. Symbols may appear as characters, setting, plot events, or specific images or objects in the story.

Book	Symbol	Meaning
The Great Gatsby	light across the bay	yearning, impossible goals
Song of Solomon	flight	freedom, escape
To Kill a Mockingbird	mockingbird	innocence, childhood, freedom
Adventures of Huckleberry Finn	the duke, the king	corruption, evil of the noble elite
A Raisin in the Sun	Mama's flower	hope for the future

When trying to determine the meaning of a symbol or image in a passage, consider the overall structure of the story. Often times, the writer will work in a set of symbols, called an *aesthetic* or *design*. As such, various symbols within a work are often related. For example, in *The Great Gatsby*, Gatsby's shirts represent the luxury of the world around him, and Gatsby "wears" them without feeling any true pride in them at all. Daisy, for her part, loves the shirts only for the wealth they represent.

Practice 3: Symbolism

Read the passage below. Then answer the questions that follow.

He was quick and alert in the things in life, but only in the things, and not in the significant things. Fifty degrees below zero meant eighty degrees of frost. Such fact impressed him as being cold and uncomfortable, and that was all. It did not lead him to meditate upon his frailty as a creature of temperature, and upon man's frailty in general, able only to live within certain narrow limits of heat and cold; and from there on it did not lead him to the conjectural field of immorality and man's place in the universe. Fifty degrees below zero was to him just precisely fifty degrees below. That there should be anything more to it than that was a thought that never entered his head.

As he turned to go on, he spat speculatively. There was a sharp explosive crackle that startled him. He spat again. And again, in the air, before it could fall to the snow, the spittle crackled. He knew that at fifty below spittle crackled on the snow, but this spittle had crackled in the air. Undoubtedly it was colder than fifty below — how much colder he did not know. But the temperature did not matter. He was bound for the old claim on the left fork of Henderson Creek, where the boys were already. They had come over across the divide from the Indian Creek country, while he had come the roundabout way to take a look at the possibilities of getting out logs in the spring from the islands in the Yukon. He would be in to camp by six o'clock; a bit after dark, it was true, but the boys would be there, a fire would be going, and a hot supper would be ready. As for lunch, he pressed his hand against the protruding bundle under his jacket. It was also under his shirt, wrapped up in a handkerchief and lying against the naked skin. It was the only way to keep the biscuits from freezing. He smiled agreeably to himself as he thought of those biscuits, each cut open and sopped in bacon grease, and each enclosing a generous slice of fried bacon.

He plunged in among the big spruce trees. The trail was faint. A foot of snow had fallen since the last sled had passed over, and he was glad he was without a sled, traveling light. In fact, he carried nothing but the lunch wrapped in the handkerchief. He was surprised, however, at the cold. It certainly was cold, he concluded, as he rubbed his numb nose and cheekbones with his mittened hand. He was a warm-whiskered man, but the hair on his face did not protect the high cheek-bones and the eager nose that thrust itself aggressively into the frosty air.

– excerpted from "To Build A Fire," by Jack London

1. What do the biscuits symbolize in the story? Why do you think so?

2. What does the old claim symbolize? Why do you think so?

3. From reading the passage, do you think the prospector will survive the cold? Why or why not?

IRONY

Irony relates to theme (the message of any given story) in that many authors use irony to make a theme's emotional impact more powerful. Irony can be the difference between what is *stated* and what is *meant* (**verbal irony**), or between what is expected to happen and what actually happens (**irony of situation**).

For an example of *verbal irony*, imagine that you and some of your friends have been working on a project for school. It's been a long, hot, and tiring day. One friend suddenly announces to the group, "I've had about as much fun as I can stand for one day!" That person is using verbal irony to express a wish to go home, saying that it's been too much fun — but meaning that it's really been too much work.

An example of *irony of situation* is displayed near the end of the movie *The Empire Strikes Back*. Darth Vader, the antagonist, reveals to Luke, the protagonist and his bitter enemy, that he is Luke's father. That the two have struggled for years is ironic because they are actually family.

Building Vocabulary & Language Skills

MOOD

Any work of fiction has a feel to it, a sense of the place that comes from really engaging the description of its setting and characters. This feeling — the soul of a work, its atmosphere — is called **mood**. An author creates mood by carefully choosing the words he uses to tell the story; using point of view; and by using a description and plot development to bring a sense of location and emotion to the work. For example, in the novel *The Red Badge of Courage*, Stephen Crane uses sharp, honest descriptions of war and injury to make the reader feel both fear and sadness.

Sample of Moods				
dismal	peaceful	anxious	joyful	elated
melancholic	chaotic	mysterious	creepy	humorous

Mood is closely related to **tone**, which is the writer's attitude toward his subject. Authors convey mood through language, word choice, and story pace. The novel *Tom Sawyer* has a tone of lighthearted nostalgia and humor. Sometimes, the tone can be reversed for humorous effect. In the novel and movie *The Hitchhiker's Guide to the Galaxy*, the tone is light and carefree, even though the story is actually very dark. A more detailed description of tone appears in Chapter Four.

Practice 4: Mood

Read the following passages first for content. Then choose the mood which most closely describes the feeling you get from the passage.

1. When the voices of children are heard on the green,

    And laughing is heard on the hill,

    My heart is at rest within my breast

    And everything else is still.

    – excerpted from "Nurse's Song," by William Blake

    A. peaceful   B. mysterious   C. lonely   D. spooky

2. Whether Faith obeyed he knew not. Hardly had he spoken when he found himself amid calm night and solitude, listening to a roar of the wind which died heavily away through the forest. He staggered against the rock and felt it chill and damp; while a hanging twig that had been all on fire, besprinkled his cheek with the coldest dew.

    – excerpted from "Wakefully," by Nathaniel Hawthorne

    A. suspenseful   B. joyous   C. uneasy   D. angry

Mood can greatly influence how an audience receives the language of a certain work. One of the most widespread kinds of mood is called *satirical*, and comes from the ancient Greek art of **satire**, which is as popular now as ever.

SATIRE

Satire involves taking the forms of a type of work and holding up their faults for public amusement or ridicule. Its goal is usually to draw attention to the form's shortcomings, show how stale its material has become, or to point out how something has lost touch with modern society's feelings and experiences.

Satire was introduced in the 5th century by Greeks who wanted a subtle way of discussing political events, and regained popularity during the Age of Enlightenment. One of the most famous works of satire in literature is Jonathan Swift's *Gulliver's Travels*, which pokes fun at human nature and failings. In modern society, satire has become very complex and specialized.

Political satire, such as found on television shows like *Saturday Night Live*, makes fun of political figures and current events in the news. **Social satire**, such as *Chappelle's Show*, frequently tries to provide humor as a means of softening a comment or criticism about society. Finally, **cultural satire** makes fun of other types of entertainment. Television shows such as *The Simpsons* and *The Family Guy* make fun of traditional television situation comedies by turning their story and character elements upside down. Homer Simpson and Peter Griffin, with their selfish and bumbling attempts to get ahead in the world, *satirize* the traditional figure of the father as the stable, emotionally mature head of the family. In the *Scary Movie* series of films, modern horror movies are deflated of their shock value by reducing their stories to slapstick episodes in foolishness.

Practice 5: Satire and Modern Entertainment

Extended Response. Find five examples of satire, both political, cultural, and social. Describe on a separate sheet of paper how each one relates to modern society and how the satires made you think afterwards.

TYPES OF RESOURCE MATERIALS

DICTIONARY

A **dictionary** provides more than the meaning or meanings of a word. It also provides different forms of the word, such as the plural form or the past tense. A typical dictionary entry would tell what part of speech the word is and give a pronunciation key. Dictionaries often give examples of the word in a sentence. A dictionary entry also provides the origin of a word.

Dictionaries are increasingly popular on the Internet. Dicitonary.com is a combination dictionary, thesaurus, and encyclopedia that is free of charge to the public.

THESAURUS

A **thesaurus** provides a word's *synonyms*, words that mean basically the same thing. It may also provide *antonyms*, words that mean the opposite. A thesaurus is a useful resource text for improving both reading and writing.

Building Vocabulary & Language Skills

ENCYCLOPEDIA

Encyclopedias are sets of books with entries about all kinds of topics. The subject entries inside each volume are also arranged by the alphabet. Entries cover a wide range of subjects such as people, places, historical events, science, and technology. Encyclopedia entries are meant to inform, meaning they provide facts without opinions.

Many encyclopedias now have online editions available. The information available from these sources is similar to the book version. However, these sources have the added benefits of references or links to other multimedia sources such as video and sound clips.

Practice 6: Encyclopedia

Gathering information from the encyclopedia. Answer the following questions by searching encyclopedia articles. Write the keyword you used to search, the type of encyclopedia you used, and the page(s) where you found the information.

1. When and where were the first International World Olympics held?
2. Who was the first United States president born in Texas?
3. In the United States, what are windmills primarily used for today?
4. What is the largest marine mammal?
5. How are cars manufactured?

MAGAZINES AND JOURNALS

Magazines and journals are great resources for current information because these publications are issued weekly, semimonthly, monthly, or quarterly. Libraries classify magazines and journals as **periodicals** (published periodically — weekly, monthly, and so on).

Magazines usually offer many articles in a specific area of interest. They are written both to inform and to entertain. Magazine articles inform readers but can also express the writer's opinion.

Journals are academic magazines that have information about specific areas of study. Since experts in each field write the articles, journals are considered unbiased, reliable sources of information. Journals are not usually found in bookstores, but they can be found in libraries or ordered from the publisher.

Magazines	Journals
Newsweek	*Nature*
Time	*American History*
People	*National Geographic*
Sports Illustrated	*Discovery*
Popular Science	*Contemporary Literature*
Seventeen	*The New Yorker*

Practice 7: Magazines and Journals

Gathering information from magazines and journals. List three sources for research on the following topics: Calgary, Alberta; the effect of *El Niño* on butterfly populations; and the popularity of skateboarding. After listing the sources, write a brief summary of the information you found in your best source.

NEWSPAPERS

Newspapers provide current information on local and world events. Most newspaper articles are written to be brief and purely informative, so they do not go into as much detail as magazine articles. Newspapers have both fact-based and opinion-based writing. **News reports** present information — facts, statistics, and statements by other people. **News features** also present information but reflect a reporter's opinion and personal style.

Newspapers are organized into sections, which help make it easy to locate articles on specific topics.

Practice 8: Newspapers

Reading a News Report. Find and read a news report from a newspaper. Create a chart telling the *who, what, when, where, why,* and *how* from the report.

Reading a News Editorial. Find and read a news editorial from a newspaper. Create a two-column chart. Write the *facts* from the editorial in the first column and the *opinions* from the editorial in the second.

ONLINE SEARCH ENGINES AND KEYWORD SEARCHES

Search engines can help you find the sites you need for information. Many search engines organize their listings by headings and subheadings. This allows you to browse through them to find the perfect resource. The Internet has many different search engines to aid you in your research. Some of the most common ones are Google, Ask Jeeves, and Yahoo.

Building Vocabulary & Language Skills

Keywords are tools for finding the most useful sites in the shortest amount of time. The results of your search depend entirely on what keywords you enter. Once you have decided on a search engine to use, type in one or two keywords for the specific subject you want information about. The engine scans its listing for sites that match the keyword(s) you have entered and lists all the results. Searching by a single keyword will usually produce more results than you need, so it is important to limit your topic search by carefully choosing your keyword(s). See the list below for tips on using keywords.

Tips on Using Keywords

1. **Create a list of words** describing the type of information you are looking for in a site.
2. **Use multiple words to limit your search.** For example, instead of searching *tigers*, you could search *Siberian tigers*.
3. **To narrow your search even more**, put your keywords inside quotation marks. This type of search will show only sites that have that exact phrase.
4. **For some search engines**, you will need to use the words AND and OR to limit your search to result entries that have either both words or one of them. For example, Siberian tigers AND habitats will yield listings containing both topics. Entering the keywords Siberian tigers OR habitats will yield listings that contain each topic.

Practice 9: Internet Research

Follow these instructions to conduct Internet research.

1. Choose one of the following controversial issues as a topic.
   - downloading music from the Internet
   - the use of metal detectors and cameras in schools
   - violent video games
   - bilingual education
   - city-enforced teen curfews
   - off-shore oil drilling
   - television talk shows

2. Perform an Internet search for your chosen topic, using keywords. As you use search engines *and* databases for the research, notice and record which tool gives you the best results.

3. Print two Web pages found in your search: one page with mostly unsupported opinions and another page with well-supported facts. Be sure that each page has the URL printed at the bottom.

Chapter 1

CHOOSING THE RIGHT RESOURCE MATERIAL

As you have seen from the description of the types of resource materials, different sources present information in different ways. Research is more interesting when you read an assortment of sources. It is also important to choose the correct resource material to find your information. Before beginning research, you need to have a clear idea of what information you are looking for. Then you need to evaluate the different resource materials and decide which source will provide you the best information.

Imagine you are doing a report on hurricanes. You will find many sources that give some information about hurricanes, but not all sources will give you enough information or the kind of information you are looking for. A dictionary will give you a definition of hurricanes. You might find a magazine or newspaper article about Hurricane Katrina and the damage it caused. You could look in the encyclopedia for information about how hurricanes are formed. You might also find a Web site on the Internet that explains hurricane tracking. You have many resources to choose from. Once you have decided what you want to say about hurricanes your choices will be easier.

Regardless of what sources you choose, all your sources should be:

- reliable — written by respected authors and published in dependable books, magazines, newspapers, and so on.
- timely — current information
- suitable — appropriate to your topic

Practice 10: Resource Materials

Choose the most appropriate resource material to find the information.

1. Alex is writing an essay and needs to find another word for *boring*.
   A. dictionary B. Internet C. thesaurus D. encyclopedia

2. Michelle is working on a science project about bats.
   A. dictionary B. newspaper C. glossary D. encyclopedia

3. Anthony is reviewing words for a vocabulary test.
   A. dictionary B. Internet C. thesaurus D. encyclopedia

4. Latisha needs to bring in an example of a current event.
   A. newspaper B. encyclopedia C. journal D. thesaurus

5. Lewis is writing about a baseball player's current season.
   A. newspaper B. magazine C. journal D. dictionary

Chapter 1 Summary

Imagery is the use of clear and vivid language to convey thoughts, ideas, feelings, scenery, or emotions.

Context clues are the words and ideas surrounding a word in a text. When put together, context will reveal clues to the meaning of unfamiliar words.

Figurative language is the crafting of text in such a way that provokes a response in the audience. It may involve repetition of words or sounds, applying human traits to nonhuman objects, or drawing comparisons between two like or unlike things.

- **Personification** is the act of giving nonhuman objects or animals traits that are uniquely human.
- A **simile** is a comparison of two unlike objects using the words *like* and *as*.
- A **metaphor** is a comparison of two unlike things without using *like* or *as*.
- An **allusion** is a reference to a specific place, historical event, literary figure or work of art.

We use the word **mood** to describe the atmosphere of a story — the feeling evoked by the writer's language and plot development. Mood is closely related to **tone**, which involves the writer's attitude towards the subject material.

Symbolism is the use of images, events, or characters in a narrative for dramatic and thematic effect. **Universal symbols** are symbols that have been used so often that their meaning is instantly recognizable to anyone.

Irony is the difference between what is *stated* and what is *meant*, called **verbal irony**; or the difference between what is expected to happen and what actually happens, called **irony of situation**.

Satire is the art of holding up the flaws in people, events, or ideas to public amusement and scorn. Satire may be political, cultural or social.

Common reference sources include **dictionaries**, **thesauri**, **encyclopedias**, **magazines** and **journals**, **newspapers**, and the **Internet**.

CHAPTER 1 REVIEW

Context Clues. Read the following article. Then answer the questions that follow.

What Is Ethics, Anyway?

Ethics is a concept we hear about, but few people today stop to think what it really means. However, philosophers and statesmen since the time of Plato have **contemplated** the definition and details of ethics, which is sometimes difficult to state. Clearly, ethics is not something invented by one person or even a society but has some well-founded standards on which it is based.

Some people **equate** ethics with feelings. But being ethical is not simply following one's feelings. A criminal may "feel" robbing a person is okay, when really it is wrong and unethical to steal. Many people may identify ethics with religion, and it is true that most religions include high ethical standards and strong motivation for people to behave morally. But ethics cannot be confined only to religion, or only religious people could be ethical. There are even cases in which religious teaching and ethics clash: for example, some religions **inhibit** the rights of women, which opposes the ethical standard of basic justice.

Ethics also is not simply following laws or what is accepted by a society. The laws of civilized nations often embody ethical standards. However, unethical laws can exist. For example, laws have allowed slavery, which is unethical behavior as it takes the freedom of another human being. Therefore, laws and other conventions accepted by a society cannot be the measure for what is ethical. Doing "whatever society accepts" may be far outside the realm of ethics — Nazi Germany is an example of an ethically **debased** society.

What ethics really refers to is a system of people's moral standards and values. It's like a road map of qualities that people want to have to be "decent human beings." It is also the formal study of the standards of human behavior. Ethics relies on well-based standards of "right" (like honesty, compassion, and loyalty) and "wrong" (like stealing, murder, and fraud). Ethical standards **encompass** ideas such as respect for others, honesty, justice, doing good, and preventing harm.

1. In the context of this passage, what does the word **contemplated** mean?
   A. thought about
   B. looked at
   C. taken apart
   D. examined

2. In the context of this passage, which of the following is closest in meaning to **equate**?
   A. compare
   B. multiply
   C. balance
   D. flatten

3. In this passage, which of the following is closest in meaning to the word **inhibit**?
   A. lie about
   B. live in
   C. give in to
   D. hold back

4. Which dictionary definition of the word **debased** best applies to its use in the passage?
   A. depraved
   B. corrupt
   C. impure
   D. distorted

5. In this passage, which of the following is closest in meaning to the word **encompass**?
   A. steer
   B. include
   C. begin
   D. mean

Building Vocabulary & Language Skills

Figurative Language

6. Saying "The cat smiled at me!" is an example of
   A. alliteration.
   B. simile.
   C. personification.
   D. metaphor.

7. "You are the sunshine of my life" is an example of
   A. metaphor.
   B. simile.
   C. personification.
   D. allusion.

8. "Ugly as a homemade sandwich" is an example of
   A. metaphor.
   B. allusion.
   C. alliteration.
   D. simile.

9. "She was more beautiful than Angelie Jolie" is an example of
   A. metaphor.
   B. personification.
   C. allusion.
   D. onomatopoeia.

10. "Mary sat musing on the lamp-flame at the table
    Waiting for Warren. When she heard his step…"

    – excerpted from "The Death of the Hired Hand," by Robert Frost

    The above is an example of
    A. personification.
    B. mood.
    C. metaphor.
    D. allusion.

11. "He's got a head like a block of concrete" is an example of
    A. metaphor.
    B. simile.
    C. allusion.
    D. personification.

12. Symbols that are understood by everyone are said to be _____.
    A. multicultural
    B. one-sided
    C. status
    D. universal

13. The difference between what is clearly stated and what is meant is known as _____ irony.
    A. verbal
    B. situational
    C. metaphorical
    D. symbolic

14. Irony that involves the contrast between one's expectations and what is actually true is said to be _____ irony.
    A. verbal
    B. situational
    C. relevant
    D. approximate

Questions 15 & 16 refer to the following passage:

About the year 1727, just at the time when earthquakes were prevalent in New England, and shook many tall sinners down upon their knees, there lived near this place a meager miserly fellow of the name of Tom Walker. He had a wife as miserly as himself; they were so miserly that they even conspired to cheat each other. Whatever the woman could lay hands on she hid away: a hen could not cackle but she was on the alert to secure the new-laid egg. Her husband was continually prying about to detect her secret hoards, and many and fierce were the conflicts that took place about what ought to have been common property. They lived in a forlorn looking house that stood alone and had an air of starvation. A few straggling spavin trees, emblems of sterility, grew near it; no smoke ever curled from its chimney; no traveler stopped at its door. A miserable horse, whose ribs were as articulate as the bars of a gridiron, stalked about a field where a thin carpet of moss, scarcely covering the ragged beds of pudding stone, tantalized and balked his hunger; and sometimes he would lean his head over the fence, look piteously at the passer by, and seem to petition deliverance from this land of famine. The house and its inmates had altogether a bad name. Tom's wife was a tall termagant, fierce of temper, loud of tongue, and strong of arm. Her voice was often heard in wordy warfare with her husband; and his face sometimes showed signs that their conflicts were not confined to words. No one ventured, however, to interfere between them; the lonely wayfarer shrunk within himself at the horrid clamor and clapper clawing; eyed the den of discord askance, and hurried on his way, rejoicing, if a bachelor, in his celibacy.

– excerpted from "The Devil and Tom Walker," by Washington Irving

15. The tone of the passage is best described as
    A. upbeat and nostalgic.
    B. sad and critical.
    C. amused and content.
    D. indifferent and innocent.

16. The mood of the passage is best described as
    A. angry.
    B. tranquil.
    C. anxious.
    D. depressed.

17. Satire that attacks the government is known as _____ satire.
    A. political
    B. social
    C. lowbrow
    D. misplaced

18. Cultural satire often makes fun of _____.
    A. the social elite
    B. changes in government
    C. entertainment programs
    D. your new hairstyle

19. The various meanings of a word might best be found in a _____.
    A. dictionary
    B. newspaper
    C. encyclopedia
    D. journal

20. Which of the following is *not* an example of a reference publication?
    A. *Time* magazine
    B. *The New York Times*
    C. The latest issue of *Supergirl*
    D. *The New England Journal of Medicine*

Building Vocabulary & Language Skills

WEB SITES

http://bartleby.com/usage
Part of a massive nonprofit resource site, this section offers exhaustive materials for every aspect of writing and editing, from grammar and usage pages to links discussing style and voice. The built-in search engine allows the user to zero in on a topic.

http://www.umich.edu/~umfandsf/symbolismproject/symbolism.html/index.html
A dictionary of symbolism from the University of Michigan, this huge site includes meanings for all kinds of symbols and themes in literature. The symbols are arranged in alphabetical listings for easy access. The site also has pages about monsters and visual symbolism.

http://www.dictionary.com
As mentioned before, type a word in its engine, and this free site gives you all possible meanings and even some etymology (word history) thrown in for good measure. Look closely for the small function keys beneath the engine blank that allow you to toggle between dictionary, thesaurus, and encyclopedia functions.

Chapter 2
Main Ideas, Passage Analysis, and Inferences

This chapter covers the following content standards:

1.C.5	Students will summarize and paraphrase main idea and supporting details.
1.C.7	Students will make inferences and draw conclusions based on explicit and implied information from texts.
1.C.G11	Students will summarize and paraphrase expository or informational text by identifying main ideas, themes, details or procedures of the text.
1.C.G12	Students will make reasonable inferences and conclusions about the text, supporting them with accurate and implied information from texts.

It goes without saying that written passages do not create themselves, springing up out of thin air or leaping whole and finished from our minds. Writing in many respects is the act of organizing information, and writing well involves creating an organized structure to frame the ideas we present.

In this chapter, we'll pick apart the art of writing well, demonstrating how effective paragraphs (and on a larger level, effective passages and books) follow an orderly process of revealing information. Remember when reading that **details** are the information that support an idea or theme, not the idea themselves. Details embellish, add suspense or humor, and serve to create a picture in the reader's mind. If main ideas are paintings, then details are the many different paints used in the creation of the finished product.

THE MAIN IDEA

"The **main idea** or central point can be found in two different ways." That sentence is a good example of a main idea topic sentence — it says in a broad statement what this paragraph will be about. **Topic** is another word for subject.

Main ideas are *broad statements about the subject of a paragraph or passage*. Statements of details alone cannot be main ideas. Following are two ways main ideas may be found:

Main Ideas, Passage Analysis, and Inferences

1) Main ideas may be **directly stated** in a topic sentence. Topic sentences can be found in the title, the introduction, or even the beginning or ending sentence of a paragraph.

2) Authors may also show the main idea by implying, hinting, or suggesting it through details and facts, not by directly stating it. This is called an **implied main idea**.

A Directly Stated Main Idea

In a **directly stated** main idea, the basic thought to be communicated is usually found either in the beginning sentence or among the ending sentences.

The example below shows a directly stated main idea in the first sentence.

Mockingbirds are common and popular birds in the eastern and southern regions of the United States. The Mockingbird is the state bird for Arkansas, Florida, Mississippi, Tennessee, and Texas. "Mockers," as they are affectionately called, are known for their ability to mimic other birds, mammals, and insects with song and sounds. Mockingbirds often live close to human homes, nesting in ornamental hedges.

After reading the first sentence, we know that the passage is about mockingbirds. The rest of the sentences give details about how the mockingbird is common and why it is popular.

Tips for Finding a Stated Main Idea

1. **Read the title.** The main topic of the paragraph or passage is often mentioned in the title.

2. **Read the first and last sentence** of each paragraph. Most of the key words and ideas will be stated in these places.

3. **Choose the answer that is the best statement** or restatement of the paragraph or passage. Your choice should contain the key words mentioned in the title, the first sentence or the last sentence of each paragraph or passage.

4. **Always read the entire passage** to get an overview of what the author is writing.

An Implied Main Idea

An author can also give information without directly stating it, by giving clues or suggestions through the details of the passage. Read the passage below and see if you can determine the season of the year being described:

Sam's favorite part of the hike was watching the colorful leaves falling through the cold, clear air.

The season isn't stated outright, but a reader can conclude the season is autumn because of the hints given. Because the season is described with details, but without using the word "autumn," the season is *implied*.

An **implied** main idea can be found by reading more than one sentence and looking for clues. Read the passage below and try to look for the implied main idea.

Traveling the Speed of Light

In 1905, a scientist named Albert Einstein published his most famous theory: the Special Theory of Relativity. The theory states the faster an object travels, the heavier it becomes. Therefore, an object traveling at 300,000 kilometers per second, or the speed of light, would be extremely heavy. The theory also connects the speed of light with the passage of time. Einstein believed everything in the universe is connected and that traveling the speed of light would slow time down. Light is the fastest thing known to science. If scientists can figure out a way to travel the speed of light, they would be able to discover whether the theory is true or false. Most scientists consider Albert Einstein a genius.

What is the implied main idea in this paragraph?

- A. Albert Einstein may be the smartest man who ever lived.
- B. The speed of light is still an important and puzzling concept for scientists.
- C. The speed of light is the fastest thing known to man.
- D. A rocket ship might weigh a lot more if it traveled the speed of light.

The title suggests the subject of the passage is the speed of light but does not give the reader enough information to identify the main idea. It is only after we read the *entire* paragraph that we understand the implied main idea as correctly expressed in choice B.

Tips for Determining an Implied Main Idea

1. **Read the title and first sentence.** Both will help you identify the topic of the selection.

2. **Read the entire paragraph or passage** to get a general understanding of the material.

3. **Note the facts and details in each paragraph.** Think of overall ideas they have in common.

4. **Choose the answer that summarizes all of the facts and ideas in the passage.** Confirm your choice by going back to the passage to check your evidence one more time.

Main Ideas, Passage Analysis, and Inferences

Practice 1: Main Idea in a Paragraph

Read each of the following paragraphs. Decide whether the main idea is stated or implied. If the main idea is stated, underline it. If the main idea is implied, write it in your own words in the space provided or on a separate sheet of paper. Discuss your answers with your classmates and teacher.

A Visit to Germany

In Germany, lunch is the main meal for most families. Supper is a cold meal which resembles a light American lunch. Potatoes are a favorite dish. They are cooked in many ways and with different spices. Germans also like to eat sausage, or "wurst." Fresh bread rolls are stuffed with wurst for a snack. Some kinds of food, like seafood, may be difficult to find in most restaurants in Germany, but the hundreds of varieties of wursts make meals interesting and tasty.

1. _____

A Golden Seal

In 1922, the Newbery Award became the first children's book award in the world. The award is named after an eighteenth-century English bookseller, John Newbery. In the library or media center, you can find the books that have won this important award. A bright gold seal is printed on the cover of all winning books. It is the best known and most discussed children's book award in the country.

2. _____

A Field Trip to the Museum

The students were excited. It was their first field trip of the year. As the school bus parked, the children watched a large outdoor Calder sculpture slowly turn in the breeze.

Inside the museum, century-old paintings hung on the gallery walls. One painting showed a silver teapot and pastries glimmering on a very heavy silver tray. One woman in the painting looked thoughtful, and the other woman sipped tea.

Other paintings showed mountains, boats, and flowers.

Time went by fast, and soon it was time to leave.

"Alright, children, it's time to move on," said Mr. Smith, the teacher. "So many artists, so little time. The school buses won't wait for us forever."

3. _____

Practice 2: Media Search for Main Ideas

A. Idea Exchange. On your own or in a group, look for paragraphs and passages on different subjects in newspapers, magazines, books, or the Internet. Write out the stated or implied main ideas you find on a separate sheet of paper. Bring the articles to class. Exchange only the articles with another student or group members. See if they identify the same main ideas that you did. Then share the results of your efforts with your instructor.

B. **Photo Titles.** Share photos or pictures with a partner. Then think of titles or main ideas to go with them.

C. **News Story Headlines.** Bring news stories to class. Cut out the headlines, and keep them separate. Exchange only the articles, and write your own headlines. Compare your own headlines with the original headlines.

In addition to the main idea, a passage may contain **subordinate ideas,** which are smaller than the main idea. The subordinate ideas are the topics or main ideas of paragraphs in a longer passage, while the main idea is the overall controlling idea of the entire passage. The subordinate ideas are the building blocks to the main idea.

An author usually supports ideas with **details** — the answers to the 5Ws and H: *who?, what?, where?, when?, why?, and how?* Details fill in information and explain the subordinate and main ideas.

Once the main idea is firmly set in your mind, you can begin to analyze the details that support it. There are two means of working with supporting details: **summarizing** and **paraphrasing**.

SUMMARIZING

Often when you read research materials, you may want to make some written notes but may not want to use actual quotations. In order to capture the main idea and supporting details from a source, it is very helpful to write a summary. A **summary** condenses the ideas in a source and allows you to use the ideas as a reference without having to reread the entire work each time.

The first step in writing a summary is to **read the entire article or source**. After reading through carefully, make a list of brief notes which will serve as a framework for creating the summary. **Write down the subject of the article**. Next, **write down the main idea** in complete sentence form. Then, **list the major details** and briefly **explain any unfamiliar terms or concepts**. Be sure to use quotation marks to indicate exact words or phrases from the original source.

The first sentence of your summary should include the title of the source and its author. Remember to use your own words, not phrases from the original, unless absolutely necessary.

A general rule of thumb is to try and keep any summary about one-fourth to one-third of the original length. If your original article is two pages long, a summary should be about one-half to about three-fourths of a page.

PARAPHRASING

To **paraphrase** means to show understanding in your own words. Paraphrasing conveys specific information from a resource without quoting it exactly. Paraphrasing allows a researcher to restate the information more clearly, in a way that is easier for him to understand during quick reference.

Exact quotes from source material can be directly lifted in a paraphrase, but should be set off from the rest of the paraphrase with quotation marks.

> **Example:** The following article by Robert Winters is used as a resource for a report on the history of civil rights in America. The author of the report would like to paraphrase the information in the following paragraph taken from the article:

Main Ideas, Passage Analysis, and Inferences

In recent history, however, change has come at a more rapid pace. Women have entered the workforce in greater numbers. The number of women in the workforce has increased from 38 percent in 1960 to nearly 60 percent in 1997. In particular, women have entered the white collar professional jobs at increasingly higher rates. Because of higher attendance at universities and colleges across the nation, the number of women professionals has rapidly increased. Women now enjoy full participation in the professions of law, medicine, and management. In some cases, more women are enrolled in graduate programs than men!

Two possible paraphrases from this source paragraph appear below. Read each, and decide which is a valid paraphrase and which quotes the material too closely without using quotation marks.

Paraphrase 1:

In the last four decades of the twentieth century, however, the number of women in the workforce increased by almost 50 percent. Higher education has played a key role in allowing women access to more white collar jobs. Throughout the nation, women are returning to school to receive degrees in well-paid professions (Winters 84).

Paraphrase 2:

The number of women in the workforce increased between 1960 and 2000. It rose from 38 percent to 60 percent. In particular, women are entering white collar professions in greater numbers. The number of women professionals has increased because of their attendance at universities and colleges across the nation (Winters 84).

Although both paraphrases are cited, the second one borrows many exact words and phrases from the original without setting off the cited passages with quotation marks. It is not a valid paraphrase.

Paraphrasing is a valuable tool for presenting detailed information in a simple and clear way. You have to make sure, however, that you understand the material enough to put it into your own words, without relying on the wording and phrasing of the original author.

Practice 3: Summarizing and Paraphrasing

Select a section at random from this book or from one of your textbooks. Practice summarizing it and paraprhasing it, using direct quotes and your own understanding of the material. Share your results with a classmate, teacher, or mentor.

THEME IN NONFICTION

Writing is really the act of communicating through words, so a literary work always relates a central idea that its author wants to share with his audience. This "big idea" is the **theme** communicated in the work. Theme is not the *subject* of a work but rather *the insight about that subject* that the work relates to its audience. Themes are most often messages about life or human nature. For example, many stories are written about love. A very common theme in literature is that "love always triumphs over evil."

A theme is rarely stated outright: being so obvious robs the reader of discovering its meaning on his own. Rather, the reader has to infer the theme by looking at all the details from the work. Figuring out the theme of a literary work is not always easy, but there are ways to make it less difficult:

- Look at the lessons the main character learns. Often the truth revealed to a character is the same truth the author wants revealed to the reader.
- Look at the conflicts and how they are resolved. This resolution of the conflict and how the conflicts change people or places often points towards the theme.
- Look at the title of the work. Sometimes titles have special meaning or give clues about the theme. For example, the title of Sebastian Junger's book *The Perfect Storm* makes the reader think about what could trigger a storm of such size and power to be considered "perfect." Its theme — that great and terrible forces of nature and fate sometimes converge with devastating effect — echoes throughout the title and prepares the reader for the events related in the text.

Read the passage below, excerpted (or taken in part) from Stephen Crane's novel *The Red Badge of Courage*. See if you can determine the theme from the answers provided.

> The youth gave a shriek as he confronted the thing. He was, for moments, turned to stone before it. He remained staring into the liquid looking eyes. The dead man and the living man exchanged a long look. Then the youth cautiously put one hand behind him and brought it against a tree. Leaning upon this he retreated, step by step, with his face still toward the thing. He feared that if he turned his back the body might spring up and stealthily pursue him.

The theme of the passage might best be expressed as:

A. The living and the dead form bonds of love.
B. Death should not be feared.
C. Never speak badly about the dead.
D. Confronting death can be terrifying.

The correct answer is *D*. While the first three choices contain general truths, they do not apply specifically to this passage: *A* focuses on love between the living and the dead, while the details in the passage convey fear and dread. Choice *B* ignores the fear so apparent in the description. Finally, *C* is incorrect because the youth never speaks badly about the dead person. Therefore, answer *D* is the best choice. It describes the overall message of the passage and is based wholly upon the details presented.

Tips for Finding Themes

1. **Read** the passage carefully.
2. **Think of one statement** that summarizes the overall message.
3. **Make sure the details** in the passage support your answer. Sometimes a statement may be true but not relevant to the passage.
4. **Make sure your answer** summarizes the message of the *entire* passage, not just one part.

Main Ideas, Passage Analysis, and Inferences

Practice 4: Theme

Read the following passages. Then choose the theme that best fits each one from the choices provided.

1. As the old man walked the beach at dawn, he noticed a young man ahead of him picking up starfish and flinging them into the sea. Finally, catching up with the youth, he asked why he was doing this. The young man explained that the stranded starfish would die if left until the morning sun.

    "But the beach goes on for miles, and there are millions of starfish," commented the old man. "How can your effort make any difference?"

    The young man looked at the starfish in his hand and then threw it safely in the waves. "It makes a difference to this one," he said.

    A. The morning sun will kill stranded starfish.
    B. Starfish must be saved from extinction.
    C. Saving even one life can make a difference.
    D. Don't walk on a beach with starfish on it.

2. It was market day. The narrow window of the jail looked down directly on the carts and wagons drawn up in a long line, where they had unloaded. He could see, too, and hear distinctly the clink of money as it changed hands, the busy crowd of whites and blacks shoving, pushing one another, and the haggling and swearing at the stalls. Somehow, the sound, more than anything else had done, wakened him up, — made the whole real to him. He was done with the world and the business of it. He let the tin fall, and looked out, pressing his face close to the rusty bars. How they crowded and pushed! And he, — he should never walk that pavement again!

    – excerpted from *Life in the Iron Mills*, by Rebecca Davis

    A. When the busy marketplace wakes up, everyone around must awake as well.
    B. A person who retires should move away from overcrowded and noisy areas.
    C. Freedom is often taken for granted until it is lost.
    D. Some people prefer to act; others prefer to watch.

3. Little Zack went outside and walked around the pool. There was something shiny at the bottom of the water, and when he looked closer, he fell in the pool. His sister Penny was making her little brother a sandwich when she heard the splash. Seeing that her brother was missing, Penny ran outside and saw him lying motionless at the bottom of the pool. Quickly, she jumped in and brought him to the surface. She had just taken a CPR course in her gym class and immediately started doing what she had learned, doing chest compressions and mouth-to-mouth resuscitation. Just then, Penny's mom came outside.

    "Mom, call 911!" Penny yelled. "Zack fell into the pool,"

    Penny's mom ran back inside and made the call. Suddenly, Zack spit out the water. Soon, his pulse returned to normal and he was breathing on his own. When the ambulance arrived, the paramedics were amazed that this little girl had saved her brother's life. She became a celebrity at school, and her class threw a party for her and the firemen who had taught her class CPR.

    A. Quick thinking and knowledge of CPR can save someone's life.
    B. Everyone should know how to use a phone.
    C. Becoming famous is a wonderful reward for learning new skills like CPR.
    D. People should not drop shiny objects in pools, especially with kids present.

Practice 5: More Practice with Theme

Choose a nonfiction work from your literature textbook. After reading your selection, complete the following activities on your own paper. For extra practice, read another work of nonfiction and answer questions 1 and 2 again.

1. Explain the theme from the work you selected. What message is the author trying to reveal?
2. Provide examples from the work to support the theme. How were you able to infer the theme? What details provided the clues?

Themes as they relate to fictional works will be discussed in Chapter Four.

INFERENCES

An **inference** is a kind of partnership. When reading, we make an educated guess based on both information given in a text as well as our own previous experiences and knowledge. Inferences usually involve combining these separate types of information and imagining what might happen next in the story. Inferences are based on what is *written*, what has been *hinted at already*, and the reader's past *experience* and *knowledge*.

Inferences are important because authors do not always flatly state what they want you to understand. But when you make an inference, you notice clues and combine those clues with what you already know. This is the "detective work" of reading; solving the mystery the author presents.

Main Ideas, Passage Analysis, and Inferences

Consider a time you grew bored with a story. Was it because you could figure out too easily what would happen? As long as a story remains hard to predict, we remain interested. When the story's ending is hard to believe based on our own inferences, we may grow impatient, exclaiming, "That ending doesn't make any sense!" or, "How unbelievable!" Many writing experts believe we only enjoy a story as long as it remains unpredictable. By making inferences, we take part in the story, setting up its ending in our imagination.

As an example, read the following excerpt from Richard Connell's short story "The Most Dangerous Game." See if you can infer what happens to the character, Rainsford:

excerpt from "The Most Dangerous Game"

Rainsford sprang up and moved quickly to the rail, mystified. He strained his eyes in the direction from which the reports had come, but it was like trying to see through a blanket. He leaped upon the rail and balanced himself there, to get greater elevation; his pipe, striking a rope, was knocked from his mouth. He lunged for it; a short, hoarse cry came from his lips as he realized he had reached too far and had lost his balance. The cry was pinched off short as the blood-warm waters of the Caribbean Sea closed over his head.

Did you guess that this character has fallen into the sea? You can correctly *infer* that he has fallen over the side of a ship because ropes and rails are common items on a ship and because the narrator says the warm water "closed over his head."

Making inferences involves carefully reading a passage and making connections among all the pieces of information. As a reader, you also have to notice what is implied or hinted at and use your prior knowledge to fill in any gaps. Being able to provoke inferences is one of the most important skills a writer can develop, because through inferences the reader comes to engage himself in the story. And if we as readers don't engage ourselves, we don't care. If we don't care, we probably won't keep reading.

Inferences can also be used in nonfiction. Look at the following online job posting. Read carefully — several times if you feel the need — and answer the two questions which follow.

Wanted

Cheerful person to work at Medical City Austin Hospital. Some experience selling helpful, but not necessary. Uniform provided. Come help us brighten our patients' days. Apply at Flower Power, 1235 Med Center Way, Austin. People with allergies to pollen need not apply.

---

1. The person who takes this job will probably work as a

    A. nurse.
    B. receptionist.
    C. salesperson in the flower shop.
    D. brain surgeon.

2. What details give hints about what kind of job this is?

You should be able to infer that the job opening is for a salesperson in the hospital flower shop. Information concerning selling as well as the name of the shop, Flower Power, are the give-away hints that lead us to the inference. That those with an allergy to pollen need not apply is another, slightly less obvious clue. The mention of a uniform, patients, and the address apply to other jobs, but the inference clues truly lead us to the correct conclusion.

Think of the mystery provided by the job posting above and the eventual correct conclusion provided by the explanation. Would you feel disappointed if the correct answer turned out to be brain surgeon or nurse? Would you feel cheated? Why or why not?

Practice 6: Inference

The following passage is excerpted from Mark Twain's *The Adventures of Huckleberry Finn*. The narrator, Huck, has been taken in by a widow and forced to live away from his beloved Missouri wilderness. Choose the inference that best fits the answers to the questions provided.

> At first I hated the school, but by-and-by I got so I could stand it. Whenever I got uncommon tired I played hooky, and the hiding [whipping] I got next day done me good and cheered me up. So the longer I went to school the easier it got to be. I was getting sort of used to the widow's ways, too, and they warn't so raspy on me. Living in a house and sleeping in a bed pulled on me pretty tight, mostly, but before the cold weather I used to slide out and sleep in the woods, sometimes, and so that was a rest to me. I like the old ways best, but I was getting so I liked the new ones, too, a little bit. The widow said I was coming along slow but sure, and doing very satisfactory. She said she warn't ashamed of me.

1. Based on the passage, we can infer that
    A. Huck likes going to school.
    B. Huck is struggling with his new life.
    C. the widow is wealthy.
    D. Huck enjoys sleeping in a bed.

2. The passage suggests that Huck is
    A. a good student.
    B. not used to living indoors.
    C. angry.
    D. trying to earn a living.

3. Which of the following events likely happened?
    A. Huck has run away from home.
    B. Huck has gone to live in a new place.
    C. The widow has made life easy on Huck.
    D. A fire has burned Huck's house down.

Main Ideas, Passage Analysis, and Inferences

4. Which of the following inferences can be made based on the passage?
   A. The widow is strict but cares about Huck.
   B. Huck once lived in a treehouse.
   C. The widow is a retired schoolteacher.
   D. Huck will become a good student.

Practice 7: More Inferences

The next time you read a story or watch a movie or television program, pay close attention to your thoughts. At what point in the story do you begin to make inferences? Are your expectations met, and if not, how does that make you feel? Pay close attention to the clues given and see how well they match your own inferences about what is happening.

CHAPTER 2 SUMMARY

The **main idea** states in a broad statement what a given paragraph will be about. Main ideas are either **directly stated**, meaning they are put down in no uncertain terms within the paragraph or passage; or **implied**, in which their appearance is more subtle, often coming together only through careful arrangement of details.

Details are the information used to support the main idea of a passage. They can be used to provoke emotion, thought, memory, or feeling within a reader. Details often include "the five Ws": who, what, where, when, why.

Summarizing information involves writing a brief explanation; **paraphrasing** involves rewriting a paragraph or passage in your own words, without changing its **meaning**.

The **theme** of a passage is its "big idea" or unifying thought. It is the message the author wants you to take away from his work.

We make **inferences** based on our own thoughts about what we read, joined to our own experiences and knowledge.

Main Ideas, Passage Analysis, and Inferences

CHAPTER 2 REVIEW

1. Main ideas that are stated outright in the passage are known as _____ main ideas.
   A. inferred	B. implied	C. directly-stated	D. factual

2. Main ideas that must be discovered through careful reading of details are known as _____ main ideas.
   A. implied	B. direct	C. inferred	D. paraphrased

3. _____ are information used to reinforce a main idea.
   A. Details	B. Summaries	C. Paraphrases	D. Inferences

For questions 4 and 5, read the following paragraphs and answer the questions that follow:

The Airborn Attack on Rabies

Rabies had been a severe threat in southern Texas during the 1990s until the Texas Department of Health Zoonosis Control Division developed a way to vaccinate wildlife affected by the disease. Planes have dropped dog food and fish baits containing a rabies vaccination hoping to curb the spread of the disease to domestic animals. According to a report from the Texas Department of Health, this outbreak caused 667 cases of canine rabies and two human deaths in 1995. Although the disease has always been found among wild animals, the real problem is that fewer than half of all dogs, cats, and farm animals, those most likely to have contact with humans, have been vaccinated against it. Since the disease is spread from wildlife through domestic animals to humans, the Texas Department of Health has decided to go to the root of the problem from the air.

4. Based on this passage, what is the main idea of the paragraph?
   A. Rabies is a very contagious and dangerous disease.
   B. Texas has found a way to curb the spread of rabies by vaccinating its wildlife.
   C. Few domestic animals have been vaccinated against rabies.
   D. Domestic animals have more contact with humans than do wild animals.

East Meets West

(1) It is amazing how the Japanese have retained their cultural heritage while simultaneously integrating many parts of Western culture. One of the most popular adaptations is the style of dress. Many Japanese today wear "Western" style clothing such as business suits, active wear, jeans, and T-shirts. Traditional clothing is often reserved for special occasions.

(2) Many Japanese also have adopted Western furnishings into their homes. It is not unusual to have a completely westernized home with only one traditional Japanese room. Western influences can be seen throughout Japanese popular culture, such as fast-food restaurants, music, and the movies.

(3) The Japanese also have more time to devote to leisure. Surveys show that spending time with family, friends, home improvement, shopping, and gardening form the mainstream of leisure, together with sports and travel. The number of Japanese making overseas trips has increased notably in recent years. Domestic travel, picnics, hiking, and cultural events rank high among favorite activities.

(4) Japan is a land with a vibrant and fascinating history, varied culture, traditions, and customs that are hundreds of years old, yet segments of its society and economy are as new as the microchips in a personal computer.

5. The best statement of the main idea is which of the following?
   A. In Japan, you will find evidence of both traditional customs and culture as well as examples of Western-style adaptations.
   B. In Japan, jeans, fast-food, picnics, and personal computers are very popular.
   C. In Japan, people always enjoy traveling overseas and shopping.
   D. Japanese and American culture are so similar that you really cannot tell the difference between them.

6. Rewriting information from a source in your own words is called _____.
   A. cheating   B. paraphrasing   C. summarizing   D. research

7. Condensing information to make it easier to read is called _____.
   A. summarizing   B. paraphrasing   C. the main idea   D. inferring

8. Write a summary for the following passage:

excerpt from "Walking" by Henry David Thoreau

My vicinity affords many good walks; and though for so many years I have walked almost every day, and sometimes for several days together, I have not yet exhausted them. An absolutely new prospect is a great happiness, and I can still get this any afternoon. Two or three hours' walking will carry me to as strange a country as I expect ever to see. A single farmhouse which I had not seen before is sometimes as good as the dominions of the King of Dahomey. There is in fact a sort of harmony discoverable between the capabilities of the landscape within a circle of ten miles' radius, or the limits of an afternoon walk, and the threescore years and ten of human life. It will never become quite familiar to you.

...I can easily walk ten, fifteen, twenty, any number of miles, commencing at my own door, without going by any house, without crossing a road except where the fox and mink do: first along by the river, and then the brook, and then the meadow and the woodside. There are square miles in my vicinity which have no inhabitant. From many a hill I can see civilization and the abodes of man afar. The farmers and their works are scarcely more obvious than woodchucks and their burrows. Man and his affairs, church and state and school, trade and commerce, and manufactures and agriculture, even politics, the most alarming of them all, I am pleased to see how little space they occupy in the landscape. Politics is but a narrow field, and that still narrower highway yonder leads to it. I sometimes direct the traveller thither. If you would go into the political world, follow the great road, follow that market-man, keep his dust in your eyes, and it will lead you straight to it; for it, too, has its place merely, and does not occupy all space. I pass from it as from a beanfield into the forest, and it is forgotten. In one half-hour I can walk off to some portion of the earth's surface where a man does not stand from one year's end to another, and there, politics are not, for they are but as the cigar-smoke of man.

9. Paraphrase the following passage:

Why You Need a Compost Pile in Your Backyard
by Ben Spiffelmeyer

The home compost pile is an efficient mulch factory, and mulch is very valuable in the garden. Your garden benefits in many ways: it looks more attractive, weeds are smothered out, moisture is conserved, and soil temperature fluctuations are reduced. Mulch can also be used to disguise unsightly surfaces or bare dirt in the garden.

Since the process of plant matter decomposing into mulch takes time, avid gardeners usually have several compost piles "cooking" at various stages of decomposition. Rather than putting grass clippings and leaves in the landfill, they can be composted and recycled back into your garden as mulch. Chemical fertilizers applied to the lawn and absorbed by the grass then get recycled through the reuse of the grass clippings.  A compost pile also benefits from some types of kitchen waste such as potato and carrot peelings, coffee grounds, and egg shells. Putting those items in a compost pile is more ecologically friendly than sending them to the sewage system by way of the garbage disposal or making them a part of the local landfill.

A healthy compost pile attracts worms, and the more worms the better. Some gardeners new to composting are distressed by the appearance of earthworms in their compost pile. Gardeners, you need to realize that those worms are very beneficial to the health of your garden soil because of their assistance in the decomposition of the vegetable and paper material. Remember, anything that is added to the compost pile becomes worm food.

Once your compost pile has matured and the bottom layer has actually turned to mulch, you can begin to reap the rewards of your work. To use the mulch as an effective weed barrier, cover the soil with a three-page layer of newspapers and then apply a layer of mulch. The bottom of the layer of mulch is constantly decomposing and adding organic matter to your soil. If the appearance of the rough, unattractive mulch is not what you want to see, it can be covered with a more aesthetically pleasing top layer of pine straw. It is allowable to put a layer of mulch up to four inches deep, but be sure to keep it pulled back a few inches from the base of plants to avoid stem and crown rot which will kill your plants.

A backyard compost pile will take some work and patience, but the rewards and benefits are well worth the time you will invest in getting it started and maintaining it.

Chapter 2

Question 10 refers to the following passage:

excerpt from *My Antonia,* by Willa Cather

I sat down in the middle of the garden, where snakes could scarcely approach unseen, and leaned my back against a warm yellow pumpkin. There were some ground-cherry bushes growing along the furrows, full of fruit. I turned back the papery triangular sheaths that protected the berries and ate a few. All about me giant grasshoppers, twice as big as any I had ever seen, were doing acrobatic feats among the dried vines. The gophers scurried up and down the ploughed ground. There in the sheltered draw-bottom the wind did not blow very hard, but I could hear it singing its humming tune up on the level, and I could see the tall grasses wave. The earth was warm under me, and warm as I crumbled it through my fingers. Queer little red bugs came out and moved in slow squadrons around me. Their backs were polished vermilion, with black spots. I kept as still as I could. Nothing happened. I did not expect anything to happen. I was something that lay under the sun and felt it, like the pumpkins, and I did not want to be anything more. I was entirely happy. Perhaps we feel like that when we die and become a part of something entire, whether it is sun and air, or goodness and knowledge. At any rate, that is happiness; to be dissolved into something complete and great. When it comes to one, it comes as naturally as sleep.

10. Which of the following statements is the best theme for the passage?
    A. Nature is to be feared.
    B. Truth is found in little things you discover around you.
    C. Happiness is difficult to find.
    D. Feeling you are a part of nature brings happiness.

11. **Short Answer.** Pick three short stories or books you have read in class this year. On a separate sheet of paper, write down any directly stated themes in one column. Then, in two more columns, see if you can identify any implied themes. Do the works you've selected relate any universal themes to the audience? If so, how?

Questions 12 – 15 refer to the following passage:

excerpt from *The Red Badge of Courage* by Stephen Crane

Turning his head swiftly, the youth saw his friend, Jim, running in a staggering and stumbling way toward a little clump of bushes. His heart seemed to wrench itself almost free from his body at this sight. His friend made a noise of pain. He and the tattered man began a pursuit. There was a singular race.

When he overtook the tall soldier, he began to plead with all the words he could find. "Jim — Jim — what are yeh doing — what makes yeh do this way — yeh'll hurt yerself."

The same purpose was in the tall soldier's face. He protested in a dulled way, keeping his eyes fastened on the mystic place of his intentions.

Main Ideas, Passage Analysis, and Inferences

"No — no — don't touch me — leave me be — leave me be —"

The youth, aghast and filled with wonder at the tall soldier, began quaveringly to question him. "Where yeh goin', Jim? What yeh thinking about? Where yeh going? Tell me, won't yeh, Jim?"

The tall soldier faced about as upon relentless pursuers. In his eyes there was a great appeal. "Leave me be, can't yeh? Leave me be fer a minute."

12. Based on the passage, we can infer that
    A. the tall soldier has been wounded.
    B. the youth has been injured.
    C. the youth is trying to arrest the tall soldier.
    D. the tall soldier wants to walk home.

13. The passage suggests that
    A. the youth is angry with the tall soldier.
    B. the youth is trying to bring the tall soldier back to the front.
    C. the youth is indifferent toward the tall soldier.
    D. the youth is trying to help the tall soldier.

14. Which of the following events is likely to happen?
    A. The tall soldier will run away from his pursuers.
    B. The youth will convince the tall soldier to return to the front.
    C. The tall soldier will tell the youth what he is thinking.
    D. There is not enough information to make a determination.

15. Based on the passage, which of the following generalizations is the most likely theme of the passage?
    A. One soldier out of three will be injured in a war.
    B. Tall soldiers are more likely to survive a war.
    C. We must choose whether or not to respond to another person.
    D. Tall soldiers are more helpful than young soldiers.

WEB SITES

http://www.rhlschool.com/read6n4.htm

This web site published by RHL School, a free online Reading learning resource center, offers tips on targeting the main idea of a passage. Simple and quick to use.

http://www.mantex.co.uk/samples/summary.htm

A British web site with step-by-step instructions on how to summarize information, with examples. Also includes easy tips on materials to use and ways to read for summary purposes.

http://owl.english.purdue.edu/handouts/research/r_paraphr.html

From the mighty English labs at Purdue University, this page offers help and guidance on paraphrasing, including its benefits and how to avoid lapsing into plagiarism. As always from the Purdue site, this is an excellent resource.

Main Ideas, Passage Analysis, and Inferences

Chapter 3
Argument, Audience, Purpose, and Credibility

This chapter covers the following content standards:

1.C.6 and 1.C.9	Students will trace the logical development of an author's argument, point of view or perspective and evaluate the adequacy and appropriateness of the author's evidence in a persuasive text. Students will identify, understand and explain the various types of fallacies in logic.
1.C.8	Students will evaluate clarity and accuracy of information, as well as the credibility of sources.
1.D.4	Students will evaluate the impact of an author's decisions regarding word choice, point of view, style, and literary elements.
1.D.7	Students will evaluate a literary selection from several critical perspectives.

Literature and writing — much like life — is filled with struggle. Writers want to make a point, and most use arguments to try and convince the reader to agree with the author's point of view, opinions, or philosophies.

In this chapter, we'll examine how to dissect arguments and test their **logic**, or their reasoning. We'll also examine an author's **credibility** — his qualifications to speak on a certain subject. As we learn to question an author's motivations, and we'll look at **style** and **clarity**, two of the most effective techniques available to the persuasive writer. Finally, we'll measure ways to gauge and enhance **audience awareness.**

WHY WE WRITE

Most people do a certain amount of **personal writing**. Some people keep diaries or travel journals when they go on vacations. Another type of personal writing is sometimes called **social writing**. Letters and thank you notes to friends or family, in either print or via email, are considered social writing.

Business writing is an important though frequently under-appreciated skill. In today's information-driven economy, virtually every business or career requires some degree of skill in writing. As an employee, you might need to record procedures for a new employee or to explain processes for employees in another location. Business writing can convey basic information, or it can be used to document technical information.

Argument, Audience, Purpose, and Credibility

Any piece of good writing has certain elements in common. A **clear focus and voice, solid evidence**, and **an awareness for the audience** are all crucial to writing well.

AUTHOR'S PURPOSE

Everyone who writes is an author to an extent, and each piece of writing may be done for different reasons. Often, the author's **purpose** is revealed in the way he writes. See if you can determine the purpose in each of the following two paragraphs.

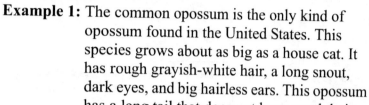

Example 1: The common opossum is the only kind of opossum found in the United States. This species grows about as big as a house cat. It has rough grayish-white hair, a long snout, dark eyes, and big hairless ears. This opossum has a long tail that does not have much hair on it. The animal can hang upside down by wrapping its tail around a tree branch. Its teeth and claws are sharp.

Example 2: One warm fall evening, our son Tom went out to the garage to feed the cat. Suddenly we heard him yell out, "A rat! A Texas-sized rat!" His older brother Joe went to investigate and reported back, "Sure enough, Mom. It is a rat!" Finally, Barb and I went to look at this "rat." When the little critter turned around to see us gawking, we realized that it was an opossum... a fat opossum.

Both paragraphs discuss opossums, but they do so in very different ways. Example 1 provides basic information about the physical features of one type of opossum. Though the paragraph contains descriptive words, there is no dialogue and no action. This paragraph would fit well in a science textbook. The purpose is to *inform*.

Example 2, on the other hand, describes characters and events with expressive words and interesting dialogue. It is part of a brief story about a surprising and funny event. Perhaps you would find it in a book of short stories. The purpose is to *entertain*.

Examples 1 and 2 show how authors can use a different style of writing depending on the purpose for their writing. The words they choose reflect a certain feeling or intent. When reading, try to determine the author's purpose, as this will enhance your comprehension. In addition, being clear about your purpose as an author will make your writing more effective.

Author's Purposes

Purpose	Definition	Reading Selection
To inform	To present facts and details	"Ocean Fishes"
To entertain	To amuse or offer enjoyment	"How My Cat Turned Purple"
To persuade	To urge action on an issue	"Raise Penalties for Polluters"
To explain why	To provide clear reasons	"How Plants Grow"
To instruct	To teach concepts and facts	"Mastering Exponents"
To create suspense	To convey uncertainty	"Will Tom Win the Race?"
To motivate	To inspire to act	"You Can Make a Difference!"
To cause doubt	To be skeptical	"Are Adults Responsible?"
To introduce a character	To describe a person's traits	"First Look at Captain Nemo"
To create a mood	To establish atmosphere	"Gloom in the House of Usher"
To relate an adventure	To tell an exciting story	"Lost in a Cave"
To share a personal experience	To tell about an event in your life	"The Time I Learned to Share"
To describe feelings	To communicate emotions through words	"When My Dog Died"

Practice 1: Purpose

Based on the list above, identify the purpose for the following reading passages. Then discuss your choices with your class or with your instructor.

1.
   excerpt from *The Red Badge of Courage,* by Stephen Crane

   The fire crackled musically. From it swelled light smoke. Overhead the foliage moved softly. The leaves, with their faces turned toward the blaze, were colored shifting hues of silver, often edged with red. Far off to the right, through a window in the forest, could be seen a handful of stars lying, like glittering pebbles, on the black level of the night.

   A. to describe an event
   B. to create a mood
   C. to persuade
   D. to instruct

2. Columbus' own successful voyage in 1492 prompted a papal bull dividing the globe between rivals Spain and Portugal. But the Portuguese protested that the pope's line left them too little Atlantic sea room for their voyages to India. The line was shifted 270 leagues westward in 1494 by the Treaty of Tordesillas. Thus, wittingly or not, the Portuguese gained Brazil and gave their language to more than half the people of South America.

   A. to introduce a character
   B. to describe feelings
   C. to relate an adventure
   D. to instruct

- Argument, Audience, Purpose, and Credibility

ARGUMENTS

An **argument** is a statement of reasons for or against something. While the best arguments will have solid **evidence**, some will use **fallacies** and **generalizations**.

Evidence is clear, provable information. It's similar to the kind of information that you see in TV courtrooms: fingerprints, shoe prints, and DNA reports. Pieces of evidence, like the identification of fingerprints, can be proven to be *true* and *correct*.

Unfortunately, a writer's eagerness to persuade his audience, no matter his intentions, frequently cause him to bend or falsify his facts. Such bad arguments can be based on **generalizations**, **fallacies**, or **assumptions**.

Generalizations are statements that are unclear and unprovable. One example is a very common statement made in election speeches: "Vote for me! I work for the common people. I feel your pain and frustration!" The candidate making the speech does not say exactly *how* work will be done for the common people, nor does the candidate say exactly what has been done to frustrate voters. Since the argument is not exact, it cannot be proven.

Fallacies are statements which can be proven untrue. Careful examination will reveal they cannot stand up to the "cold light of truth" and be trusted. Fallacies often bear such close resemblance to the truth that their trustworthiness seems almost absolute. As such, you have to be very careful, listening and reading accurately, to catch fallacies as they present themselves.

Assumptions are broad beliefs that are accepted as true without actual proof. When a politician says "I work for the common people," he assumes that he is talking to those who consider themselves part of the "common people," and will identify with what he says. Another assumption is that the crowd is frustrated.

To correct this argument, the candidate could point to exact actions, like unjust tax laws, that have frustrated voters. The candidate could also point to positive actions (provable ones) that he has taken to improve the lives of voters.

Tips for Analyzing an Argument
1. Identify the opinion or viewpoint on the issue. It will generally appear near the beginning of the selection, but it may also appear at the end as a conclusion.
2. Read the supporting reasons or evidence for the opinion.
3. Identify any assumptions the author has made.
4. Identify statements that imply additional information.
5. Decide whether the reasons or the examples of evidence support the argument.
Hint 1: A **valid argument** contains good logic, solid evidence, or clear reasons and examples.
Hint 2: A **fallacy** or **false argument** contains poor logic, weak evidence, or faulty reasons and generalizations.

Chapter 3

Practice 2: Arguments

Read the following paragraph, and answer the questions that follow:

To the Editor:

Listen up fellow students! As class president, I have a lot of demands on my time, so I'm only going to say this once. Stop drinking those supercharged caffeine sodas! During these last few weeks of exams, I've been seeing people walking around with glazed eyes and jittery hands until they pick up their first morning soda and drink it quickly. Then they show the worst signs of the hyper, ill-tempered caffeine and sugar addicts. Yeah, kids say overdosing on caffeinated soda is no worse than drinking coffee, but is it any better? Why go for the same legal high that your parents go for? Seeing kids hugging the cola machine all day is as sad as seeing the lines of hollow-eyed adults lined up at the coffee shop. As your class president, I'm concerned about the image we show to the other schools. The ads on commercials make drinking sodas look so cool. Not true. As students, we need to get smarter and to declare war on the dangers of caffeine and the chemicals in sodas. The truth is that researchers say soda is making you sick, much sicker than you think. Healthy people never need caffeine. Sure, we all could use help to stay awake sometimes, but chemical and sugar drinks cannot be the answer. If you are one of the unfortunate people who can't make it through the day without at least five super-sized colas, take a good look in the mirror and ask yourself how you feel. Then ask yourself "why?" The answer may lie inside your next bottle of soda.

Colin Brown, Sophomore Class President

1. Which one of the following statements BEST describes the writer's argument?

   A. The writer states that healthy people never need caffeine, a generalization.
   B. The writer develops a valid (good) argument.
   C. The writer uses statistics to support the argument.
   D. The writer uses generalizations in developing this weak argument.

Practice 3: Reading Arguments

For each passage below, list the **assumptions** the author has made about the situation or audience, and decide if the author has used **evidence** or **generalizations**.

1. ### Why I Am in Favor of Year-Round School

   There are a lot of rumors going around the school and all over the county, about the school board considering a change to a year-round school schedule. I am a student here at Central Valley High School, and I think we students should give this a chance to work. It could be a good change for students. I don't know the exact plan the school board is proposing, but here are some reasons that I think this could be a change for the better: (1) If we are in school more days each year and semester, we could probably have shorter school days. (2) Even though we wouldn't have a long summer vacation, we could have more short vacations during the year. Everybody gets burned out after 8 or 10 weeks, so we could have a week off at mid-semester and then maybe two weeks or more between the semesters. (3) Every fall, we waste a lot of time reviewing what we did the year before. If we didn't have such a

Argument, Audience, Purpose, and Credibility

long break, we wouldn't have to do that. Before you jump to the conclusion that year-round school is all bad, take some time to think it over, and there may be some real benefits to the students.

2. **excerpt from *Shams* by John S. Draper**

   Doctors are a regular set of humbugs, and most of them are quacks. They go off to school and learn some recipes for curing diseases. Then they manage in one way or another to get the teachers in their school to give them a certificate, and then they'll go out into some town or village or city and rent one space over a drugstore and get some bones and an old skull and a lot of books and spread them around the room and call it an office. They put up a sign and call themselves doctors. Then they guess what your illness is and collect their money.

3. I love the Ozark Mountains. After living in the city for forty years, I appreciate the simple courtesies. People wave to you, and you're expected to wave back. John Simmons, a long-time resident, tells me that when he didn't wave to his neighbor, Nancy Conner, she told him she was not sure she would wave to him again. People in the Ozarks also help each other out. When my cousin's house burned down, all the neighbors pitched in and built a new house for her in eight weeks. When someone dies, the neighbors cook, clean, and care for the family members of the deceased. I'm glad I answered the call of the mountains urging me to return to my roots.

VOICE, CLARITY & STYLE

Writers write for different purposes: to amuse, entertain, move to action, anger, and many more. For every human experience, there is probably something written that encourages it. But the language used in writing varies according to its subject matter, and, as readers, we are more adept at ever at detecting faulty or imperfect language.

A writer's unique way of conveying the point he wants to express is called his **style**. A writer can use various types of style such as humorous, serious, short, simple, complex or persuasive. The extent to which he can shape that style to be understood and appreciated by his audience determines the **clarity** of his writing "**voice**." All writers have a style and voice, from the giants of literature to the most inexperienced writing student. You have a voice yourself, and, though you are just beginning to develop it, it is your own.

In nonfiction, a neutral voice is usually used in the presentation of facts. The writer is careful not to let emotion cloud the presentation of events. In persuasive writing such as commentaries and editorials, the author will sometimes adopt a **tone** (or attitude towards his subject) of heated anger, cool aloofness, or even detached amusement.

When you read, pay careful attention to the voice being used. If the wording and phrasing seem inappropriate, the theme of the work is that much harder to take seriously.

Practice 4: Voice and Style

Make a list of five writers, fiction or nonfiction, whose style you admire. Write them down on a separate sheet of paper and describe what about their style that you like. Share your results with a classmate or teacher.

CREDENTIALS: BUILDING TRUST

Would you trust a plumber to repair the engine block in a car? Would you call a mechanic to fix a leaky faucet? Probably not. Because our world is too complex for one person to know how to do everything, we often turn to experts to perform a service or offer their wisdom.

Credentials are the evidence — the facts — that someone is an expert on a given subject. Credentials may include an advanced university degree, formal training in a special area or study, or many years of work experience.

The word *credential* comes from the Latin word for "authority" and "trust." Such words as *credo*, *credibility*, and *credit* also come from that same root. Likewise, the word *incredible* means "hard to be believed." So, think of credentials as the proof someone knows their subject. Their credentials prove to the world that there is trust behind their work and opinion because their education and/or experience deserve it.

When determining whether or not to believe the evidence put forth by an essay or argument, considering the author's credentials is a great place to start. Credentials, like credit, allow them to put forth ideas and opinions, like writing checks that are guaranteed by a bank.

Credentials normally appear at the bottom of an essay or article. Many times, you will see them in italics underneath the end of an essay on the editorials page. An example of a piece written by an expert on space exploration might look like this:

Dr. John Kelvin is a Professor of Astrophysics at the University of Minnesota. He served as a flight engineer on the Apollo space missions from 1972 – 74.

Dr. Kelvin's position teaching a relevant subject — astrophysics — at a university qualifies him to talk about space travel. His work with the Apollo space program is further proof that he is an expert. Therefore, a reader can reasonably expect his opinion about space exploration to be accurate or at least well-informed.

Credentials will also appear near the preface of a textbook or book. As an example, credentials for the writers and editor of this book appear in the Preface. The editor's and authors' credentials tell you of their experience in teaching English and writing. They also list their degrees related to writing about English and any important work-related experience. The other writers all have experience in writing and teaching as well. If they didn't, you might not trust the ideas presented in this book.

Argument, Audience, Purpose, and Credibility

Practice 5: Credentials

Match the credentials with the persuasive writing of which they are most qualified to speak.

1. Doctor of Psychology _____
2. Construction foreman _____
3. Veteran Navy officer _____
4. Mother of three children _____
5. Author of three books on education _____

A. "We need more battleships in the fleet."
B. "The local school lunch program is too expensive."
C. "Dealing with your fear of tall buildings"
D. "Why detention doesn't work"
E. "The new city council building is unsafe."

Practice 6: More About Credentials

Go through your textbooks, looking at the credentials in the various prefaces and introductions. Notice how each one gives the qualifications of its authors to present evidence. Did you find any examples of bad credentials or situations where credentials seemed weak? Why or why not? Write down your answers on a separate sheet of paper.

AGENDA

Most publications will offer experienced, knowledgeable experts. It is how they are able to offer their product to potential users. Sometimes, however, you will see persuasive writing by people who claim credentials but do not expect you to look much further than a title. They will try to convince you they have an expert opinion, simply because of their degree or title. These are very similar to people who use opinion pieces to help themselves, masking their goals as persuasive speaking. Such people are said to have an **agenda** — an objective they want their writing to accomplish. Their work should be met with great skepticism.

Be careful of noticing those who claim to be doctors or experts, when their positions or credentials do not match the field they are speaking of. An example would be a military commander offering medical advice or vice versa. When encountering this situation, you must decide for yourself whether to believe the evidence presented.

Practice 7: Grading Credentials

Below are the titles of five opinion pieces and the credentials of their authors. If the credentials qualify the author as an expert, write "Pass." If they do not, write "Fail." Use a separate sheet of paper to record your decisions.

1. "Vita-Care vitamins are the best supplement on the market" by Dr. Dean Frankle, Ph.D. of History.

2. "The teenage curfew law violates the Constitution" by Adam Tracey, former federal court judge.

3. "The holiday season is an overblown waste of money" by Ebenezer Scrooge, president of Scrooge & Marley Bank.

4. "Reading Amazing Comics will make your hair fall out," by Dave DiCarlo, president of Bang Comics.

5. "Coping with Depression After the Loss of a Loved One," by Dr. Judy Anders, clinical therapist.

MAKING INFERENCES ON BIAS AND THEME

To have **bias** is to have strong personal feelings for one side of an argument over another. Detecting whether a work of non-fiction is biased largely depends on understanding a piece's **tone**. Tone can have many forms:

Types of Tone In Nonfiction				
angry	stiff	dramatic	optimistic	sad
anxious	relaxed	fearful	pessimistic	tragic
rude	hysterical	flippant	humorous	satirical
merry	expectant	lofty	threatening	serious
frustrated	apologetic	sarcastic	sympathetic	objective

The tone of a non-fiction piece goes a long way towards helping you detect its purpose. If the tone is emotional but not straightforward, there is a good chance a piece is written to convince you of something. An example of such writing would be an editorial in the newspaper or commentary on a television news network. If it is unemotional and deliberate, it is probably only trying to relate a set of facts — for example, a report on television. Tone as it applies to fiction will be discussed in the next chapter.

In nonfiction, there are essentially two kinds of essays. Both have completely different purposes:

Expository writing: Written to explain a story or set of circumstances. Examples include newspaper and magazine articles, government documents, personal essays.

Persuasive writing: Done to convince an audience that certain ideas are true. Examples include letters to the editor, editorials, political campaign speeches and media advertisements.

For example, let's say your school board has invited students to write compositions about a plan to install security video cameras in the football stadium. These compositions can be expository or persuasive. Read these introductory examples from each composition type:

1.   Security cameras were invented almost as soon as television was invented in the 1950s. So being able to film crimes when they happen is not something that is brand new or cutting edge. Police using such cameras is a good thing. Other people would use the cameras for bad purposes, however.

Argument, Audience, Purpose, and Credibility

2. As citizens, our basic rights include the right to privacy. Filming private citizens without their consent or knowledge clearly violates that important right and the right of freedom of movement. We need to guard these rights even if we are not yet old enough to vote.

The first paragraph presents a historical view and mentions positive and negative possibilities for the cameras. It is *objective* in tone. The second introduction uses words that most people would have positive reactions to: rights, freedom, and private citizens. This passage is clearly *biased* to persuade. Understand that there is nothing debatable about the value of rights and freedoms themselves, only that the author uses them to make his point.

When reading a passage meant to present an opinion or bias indirectly, it is important to notice the language the author uses to make his point. Also notice if the author is presenting only one point of view, writing as if no other options existed.

Tips for Recognizing an Author's Bias

1. **Detect the author's opinion on the issue.** Though generally appearing near the beginning of the selection, the viewpoint may also appear at the end. It may be stated directly or become obvious through the presentation of ideas.

2. **Study the supporting information the author chooses to include.**

3. **Listen to the tone used in the passage.**
   - The tone is **objective** when the passage is written with no emotion and with facts presenting both sides fairly.
   - The tone is **biased** if the piece is obviously written with emotion or passion. The author's sympathy may also be directed at one side of the topic. In such cases, the author's tone towards the other side will often be indifferent or even critical.

4. **Look out for "jingo."** *Jingo* involves using sharply patriotic words to bully people into supporting a cause. Words such as *freedom, rights, liberty, equality, fairness, honor, community, family, justice,* and *courage* will be used to rally support to the author's side. The other side of the issue will be ignored or the language used to describe it will be negative. Examples include *irresponsibility* instead of *freedom, wants* instead of *rights, disloyalty* instead of *dissent, chaos* instead of *liberty, weakness* instead of *fairness*.

One of the most famous examples of persuasive writing in American history was written by Thomas Paine, shortly before the War of Independence. Read the passage below, watching how Paine skillfully appeals to love of family and liberty to convince others to join the rebellion against Great Britain. Especially appealing words and phrases are highlighted.

Thomas Paine

> "**These are the times that try men's souls**.
>
> The summer soldier and the sunshine patriot will, in this **crisis**, shrink from the service of their country; but **he that stands it now deserves the love and thanks of man and woman**.
>
> **Tyranny**, like **hell**, is not easily conquered; yet we have this consolation with us, that the harder the conflict, the more **glorious** the **triumph**.
>
> What we obtain too cheap, we esteem too lightly: **it is dearness only that gives every thing its value**.
>
> **Heaven** knows how to put a proper price upon its goods; and it would be strange indeed if so **celestial** an article as **freedom** should not be highly rated.
>
> Britain, with an army to enforce her **tyranny**, has declared that she has a right (not only to tax) but "To **bind** us in all cases whatsoever," and **if… [that] is not slavery, then is there not such a thing as slavery upon earth**. Even the expression is impious; for so unlimited a **power** can belong **only to God**."

– from "The Crisis No. 1"

Paine inflates the issue of taxation between England and the American colonies into an epic confrontation between good and evil. Not surprisingly, the essay was highly effective in getting colonists to rise up in revolution.

Practice 8: Biased Attitude

For each of the following examples, write two or three sentences using words that will influence or persuade the intended reader. Use your own sheet of paper.

1. Convince your mother to let your friend sleep over.

2. You want to be a vegetarian. Convince your parents that it is the healthiest choice to make.

3. Your little brother was sick last night, so you couldn't finish your homework. Persuade your teacher that everyone should have a second chance.

4. While mowing your neighbors' lawn, you ran over their flower bed. Convince them that this was good for their yards.

Practice 9: More Biased Attitude

Look through newspapers and magazines. Find three examples of biased or slanted language, and explain on your own paper the author's attitude towards the topic. Then find three examples in which the language is objective, not biased, and rewrite them in order to persuade the reader to support or agree with your position on an issue.

TELLING FACTS AND OPINIONS APART

Once you can rapidly tell a fact from opinion, the mystery of a written passage will quickly give up its secrets. Though the difference between fact and opinion is obvious, a formal definition will help us in our examination.

Put simply, **facts** state information based on observation, statistics, or research. They can be proven, as if in a court, "beyond a shadow of a doubt." All evidence is a form of fact.

Opinions express a personal viewpoint or belief about a person, place, event, or idea. With an opinion, there is always room for debate or doubt. Opinions are not necessarily the same as fallacies, generalizations, or assumptions and may in fact be deeply held beliefs. However, they are nonetheless not the same as provable facts.

Tips for Identifying Facts and Opinions

1. **Facts state information** based on observation, statistics, or research.
2. **Opinions express a personal viewpoint** or belief about a person, place, event, or idea.

   Hint 1: Opinions contain adjectives like best, worst, favorite, dishonest, fun, and so on.

   Hint 2: Opinions sometimes include phrases such as "you should," "I think," "my view," "my opinion," and so on.

Look below at some facts and opinions.

Fact:	Many vegetables contain vitamins that are essential for health.
Opinion:	Vegetables are easy to cook and delicious.
Fact:	Oprah Winfrey was born in Kosciuska, Mississippi, on January 29, 1954
Opinion:	The Oprah Winfrey Show is the best talk show on television.

The first statement about vegetables' health benefits is a **fact**. Researchers have proven that vegetables contain important vitamins and minerals. However, the sentence describing vegetables as "easy to cook and delicious" is an **opinion**. It expresses a viewpoint, since not all people think vegetables are delicious or easy to cook.

The statement about Oprah Winfrey is a **fact** because official birth records prove the place and date of her birth. The second statement is clearly an **opinion**. The phrase "the best" only describes one person's belief about the Oprah Winfrey Show.

Practice 10: Distinguishing Facts and Opinions

Read the following paragraph. See if you can distinguish the factual statements from the opinions.

1) Viola Clay is 110 years old. **2)** She has a 73-year-old grandson on Social Security. **3)** Viola is delightful, charming, and civil most of the time. **4)** At precisely 2 p.m. every day, her 92-year-old daughter Tillie visits her mother at the Forest Hills Nursing Home. **5)** It's nice that she brings some cake or cookies to share. **6)** "When she gets cranky, I know it's time to leave," her daughter says. **7)** According to birth records, Viola lived through 22 United States presidents. **8)** "My mother's older than the Titanic," Tillie tells her friends.

On the spaces below, write the sentence numbers for the fact and opinion statements.

Facts: _____ Opinions: _____

Practice 11: Finding Facts and Opinions

Go through newspapers and magazines looking at sentences that appear to be factual. Can you find evidence of opinions? Record your answers and consider why the author of the piece might feel the way his opinions suggest.

AUDIENCE AWARENESS

When you write, you must consider your audience — the person(s) who will read what you write. Unless you are writing in your journal or taking notes in class, you are always writing for a particular audience. It may be your teacher, a friend, your parents, or a manager at work. Knowing your audience gives you important information including the following:

the audience's interest:	what topics or information is of interest to the audience (so you can capture the interest of your readers).
the audience's prior knowledge:	what the audience already knows (so you don't tell the readers something they already know, and you can draw on that prior knowledge).
the audience's vocabulary:	words that the readers understand (so you don't use words that are too easy or too difficult).
what the audience needs to know:	information or explanations that you want the audience to know (so you can choose what information to include).

Argument, Audience, Purpose, and Credibility

Read the following two paragraphs, written by the same person for separate audiences. Try to develop a picture of the audience that the writer had in mind.

Example 1: Since you're in the market for a new car, I wanted to tell you about mine. My new car is the best one I've owned. It's a 2005 Puma. It's got a 5.0 L overhead cam engine with multi-port fuel injection. It can do 0–60 m.p.h. in 5 seconds. With that much engine, passing cars on the highway is a breeze, but handling corners on back roads is a little trickier than with my old pickup. I love the rush I get when I'm cruising around with my new wheels. You should consider buying one, too.

Example 2: Since you're in the market for a new car, I wanted to tell you about mine. My new car is the best one I've owned. It's a 2005 Puma. This sporty two-door is canary yellow with electric blue racing stripes and silver mag wheels. It has cordovan leather seats and a concert hall quality sound system. The sunroof is the perfect finishing touch. You should see the looks I get when I'm cruising around with my new wheels. You should consider buying one, too.

In both paragraphs, the author is telling someone about a new car, but each paragraph includes very different details about the car. Based on these differences, how would you describe the intended audience of each paragraph? What evidence is there for your description?

AUDIENCE INTEREST

How does the writer try to catch the audience's interest in each paragraph? Clearly, the first paragraph is intended for a reader who is interested in a car's power and performance. The writer describes the car's engine, as well as its speed and handling. The second paragraph, on the other hand, mentions nothing about performance. The writer assumes that the audience is concerned with appearance and style, so the description focuses on colors and high-priced options.

AUDIENCE KNOWLEDGE

What does the writer assume the audience already knows? Since the reader of the first paragraph is interested in performance, the writer assumes that the reader knows what a Puma is and that going 0-60 m.p.h. in 5 seconds is fast. The reader of the second paragraph may need the author to describe the Puma as a "sporty two-door," but the reader understands the stunning colors and fine accessories of the new car.

AUDIENCE VOCABULARY

What kinds of words will the audience be familiar with and understand easily? The writer expects the reader of the first paragraph to know technical terms like "5.0 L" and "multi-port fuel injection." While these terms may speak loudly and clearly to the reader of the first paragraph, they may mean nothing to the reader of the second paragraph who appreciates "cordovan leather" and a "concert hall quality sound system." Likewise, the reader of the first paragraph may have no use for these terms since they have nothing to do with power or performance.

Chapter 3

WHAT THE AUDIENCE SHOULD KNOW

What does the writer want the audience to know? In both paragraphs, the writer wants to share excitement about a new car purchase in order to encourage readers to purchase the same kind of car. The writer shares information that will be of interest to two kinds of audiences and that will encourage readers to purchase a Puma.

Various writing assignments or "real-life" writing situations will require you to address a particular audience, such as parents, teachers, other students, or the editor of a local newspaper. Knowing your audience and taking their interest, level of information, and vocabulary into account will make your writing more concise, appealing, and powerful.

Practice 12: Audience

For each of the following topics, describe the interest, knowledge, and vocabulary of the given audience, as well as what you think the audience should know.

1. Topic: parental advisory stickers on music CDs

    Audience: students

    Audience Interest

    Audience Knowledge

    Audience Vocabulary

    Audience Should Know

2. Topic: using lottery to fund public education

    Audience: governor of state

    Audience Interest

    Audience Knowledge

    Audience Vocabulary

    Audience Should Know

3. Topic: high salaries of professional athletes

    Audience: stadium worker

    Audience Interest

    Audience Knowledge

    Audience Vocabulary

    Audience Should Know

FORMAL AND INFORMAL LANGUAGE

You should use **formal** language for any writing assignments in school or for any standardized tests. Formal language is appropriate for business letters or letters to people who hold a particular office, such as a superintendent, mayor, or newspaper editor. **Informal** language is used when addressing friends, family members, or people with whom you feel comfortable and relaxed.

Argument, Audience, Purpose, and Credibility

Formal Language	Informal Language
Characteristics	
broader vocabulary	simple words
more complex sentence structure	simple sentences
strict attention to proper grammar	loose following of grammar rules
no slang	can use slang, depending on audience
Appropriate Uses	
written assignments for school	conversations with friends or relatives
business letters	personal letters, e-mails

Practice 13: Formal and Informal Language

Read each sentence below. Write on the line next to the sentence whether it contains formal or informal language.

1. You must consider the feelings of each person before making your decision.

2. She's never gonna go for that.

3. The Internet is a great place to get lots of cool information.

4. Yo, you wanna catch a movie Saturday night?

5. Be sure to pay us a visit while you are in town.

Chapter 3 Summary

Good writing, no matter what purpose, includes a **clear focus**, **solid evidence**, **a distinct voice**, and an **awareness for the audience**.

A writer's **purpose** — his reasons for writing — are often revealed in the way he writes.

Arguments are statements or reasons that support or oppose a person, situation, or thing. The best arguments have plenty of **facts** or provable information to back them up. Some arguments attempt to persuade readers with **fallacies** and **generalizations** - information that is misleading or outright false.

Facts contain information based on observation, statistics, or research.

Opinions express the writer's own personal views about a subject.

An author's **style** is his own way of communicating through words and language. The extent to which that style is easily understood is its degree of **clarity**.

Credentials are the evidence that someone is qualified to speak with authority on a given subject. Writing by those not qualified to offer evidence or by those who try to confuse readers by presenting irrelevant credentials is said to be part of an **agenda**.

Audience awareness involves tailoring your writing to a particular audience. Proper attention will include gauging their level of **interest**, awareness of their **prior knowledge** on a given subject, their **vocabulary**, **relevant** information, and **formal** or **informal language**.

Argument, Audience, Purpose, and Credibility

CHAPTER 3 REVIEW

For questions 1–3, identify the author's purpose from the list of choices provided:

1. Hand grippers can help give your arms those bulging biceps you're after, but only if they offer enough resistance. If you can squeeze them repeatedly for one to two minutes, and your hands don't get tired, they're too weak for you. You can keep buying stronger ones or make something at home that can do the same job.

    A. to instruct about the correct use of hand grippers
    B. to persuade people to buy the right kind of hand gripper
    C. to describe an event that happened while using hand grippers
    D. to instruct how to make hand grippers at home

2. Twelve-year-old Nadia told us a true story that seemed unbelievable about her family's journey to the United States. In Romania, her father was involved in politics on a national level, opposing the communist regime of Ceausescu. Because of death threats made against her family, Nadia and her family had to leave the country in the middle of the night. They had arranged for a boat to meet them, so they could sail the Black Sea to freedom in Bulgaria. Because they left in a hurry, they took nothing with them except their clothes. The boat never came, so they swam into the sea until a boat from Bulgaria discovered her family swimming. All of them survived the swim except for Nadia's little brother, Dimitri. From that country, they obtained refugee status and traveled to the United States. Nadia is very grateful to be living here, and after hearing her story, so are we.

    A. to instruct readers about the history of Romania
    B. to describe an event in the politics of Eastern Europe
    C. to persuade readers that democracy is a violent form of government
    D. to relate an adventure one family had while escaping oppression

3. My family came to America in 1985. No one spoke a word of English. In school, I was in an English as a Second Language class with other foreign-born children. My class was so overcrowded that it was impossible for the teacher to teach English properly. I dreaded going to school each morning because of the fear of not understanding what people were saying and the fear of being laughed at.

    –Yu-Lan (Mary) Ying, an eyewitness account about learning English

    A. to cast doubt on the wisdom of immigrating to the United States
    B. to motivate schools to spend more money on English classes
    C. to share the personal experience of a foreign-born English student
    D. to create suspense about whether English classes are useful for all students

Question 4 refers to the passage below:

...After all, there is an element in the readjustment of our financial system more important than currency, more important than gold, and that is the confidence of the people. Confidence and courage are the essentials of success in carrying out our plan.

You people must have faith; you must not be stampeded by rumors or guesses. Let us unite in banishing fear. We have provided the machinery to restore our financial system; it is up to you to support and make it work.

It is your problem no less than it is mine. Together we cannot fail.

– Franklin D. Roosevelt, 1933

4. Which of the following best describes President Roosevelt's argument?
   A. The American people will have to deal with problems themselves.
   B. Gold and currency have ruined the American sense of confidence.
   C. The only thing we have to fear is a crippling, decade-long national depression.
   D. Confidence in ourselves and our economy is more important than money.

Question 5 refers to the passage below:

Can you name the material that can be used as an antiseptic salve, a treatment for stomach ulcers, an embalming fluid, and a great tasting topping for your morning toast? No such thing, you say? Be ready to be surprised. The answer is honey! The sticky, sweet, super-saturated sugar is all those things and more. The use of honey through the ages has met both medicinal and nutritional needs. Many cultures have used honey as a balm for wounds. Now, scientists have proven that the high level of sugar in honey prevents bacterial growth, and the fluid has moisturizing properties that promote healing. Honey works as a healing balm also for stomach ulcers when eaten by itself, with some foods, or added to herbal teas. As for embalming, the Greeks, Romans, and Egyptians all used honey to preserve the corpses of their revered dead. Alexander the Great's body, it is said, was shipped in a wooden cask filled with honey to his home in Greece for burial. Some of Alexander's funeral foods were most likely prepared with honey to sweeten and to thicken them. So, tomorrow morning take a jar of sourwood honey and pour the golden drops of bee-ripened nectar on your biscuit or into your tea. You'll see that honey is truly one of nature's sweetest surprises on Earth.

5. Which of the following best describes the author's argument?
   A. You catch more flies with honey than with vinegar.
   B. Honey has a variety of purposes for which it's not well known.
   C. Scientists' research into honey has produced alarming findings.
   D. Honey is nutritious and easy to use.

6. Think of two works of fiction you have read or seen on video or television in the past year. On a separate sheet of paper, write down their style. Consider how their styles are different in describing people, places, and events. Discuss your results with friends or classmates.

Argument, Audience, Purpose, and Credibility

7. Which of the following is *not* an authentic source of credentials?
   A. formal training
   B. years of work experience
   C. opinions of friends and relatives
   D. advanced university degree

8. Credentials will usually appear in which part of a textbook?
   A. the appendix
   B. the glossary
   C. the table of contents
   D. the preface

9. A set of goals masked by persuasive writing is sometimes called a(n) _____.
   A. evidence   B. tone   C. credential   D. agenda

For Questions 10 – 14, match the credentials with the persuasive writing of which they are most qualified to discuss:

10. high school principal
11. police captain
12. registered nurse
13. award-winning movie director
14. Bible scholar

A. "The Life of St. John the Baptist"
B. "Preventing Infection in Everyday Cuts and Scrapes"
C. "Teenage Delinquency Hampers Learning."
D. "The New Curfew for High School Students Is In Everyone's Best Interest."
E. "Making A Quality Home Movie"

15. Writing that is found in a scientific report is an example of _____ language.
    A. formal   B. informal   C. persuasive   D. narrative

16. _____ is the information the audience already knows.
    A. Prior knowledge   B. Interest   C. Credential   D. Agenda

17. _____ language includes simple words, sentences, and a loose obedience to grammar.
    A. Formal   B. Informal   C. Slang   D. Technical

18. Which of the following is *not* a foundation of good writing?
    A. clear focus
    B. distinct voice
    C. solid evidence
    D. generalizations

19. Which of the following is an acceptable basis for an argument?
    A. evidence   B. generalizations   C. fallacies   D. assumptions

20. For the following topic, describe the interest, knowledge, and vocabulary of the given audience as well as what you think the audience should know.

   Topic: Parental Advisory stickers on music CDs

   Audience: concerned parents
   Audience Interest
   Audience Knowledge
   Audience Vocabulary
   Audience Should Know

WEB SITES

http://www.lib.unc.edu/instruct/evaluate/web/bias.html

A web site from the University of North Carolina that offers help with spotting bias online. Free to the public and with helpful links to similar sites. Includes easy sample graphics that provide interactive exercises in spotting credentials and bias.

http://www.library.cornell.edu/olinuris/ref/research/webeval.html

Similar to the web site listed above, Cornell's evaluation guide lists criteria and methods for tracking down bias and credentials through context clues and steady investigation.

http://www.powa.org/

This site features long discussions on how to write essays and some tips on revising and editing work. Clearly written and organized, this is a good source for those who want a review of the composition process.

Argument, Audience, Purpose, and Credibility

Chapter 4
Literary Elements

This chapter covers the following content standards:

1.D.4	Students will evaluate the impact of an author's decisions regarding word choice, point of view, style, and literary elements.
1.D.6 and I.D.10	Students will analyze and evaluate the relationship between and among elements of literature, character, setting, plot, tone, symbolism, rising action, climax, falling action, point of view, theme and conflict/resolution. Students will interpret the effect of literary devices.
I.D.14	Students will respond to literature using ideas and details from the text to support reactions and make literary connections.

Literature in its current form has existed for more than six centuries, but the tools of conveying a story haven't greatly changed. These "working parts" are found in all literary genres and have become a fundamental part of the narrative process.

When you read a novel or short story, your close mental involvement will reveal its basic elements by recognizing the parts that make a work complete and whole. This is the real power of critical reading and the first step in becoming a reading scholar. Rather than passively absorbing a story, your relationship with the work becomes a kind of "mind reading" because you're able to see the author's goals — and faults — in presenting a story.

For a book or short story — or movie or television program — to function well, several elements must work together in perfect harmony, creating an effect that is greater than themselves alone. These parts include:

Theme:	the story's "message"
Setting:	the story's "background"
Characterization:	the story's "cast"
Point of View	the writer's perspective
Tone	the writer's attitude toward the characters and events
Plot:	the story's "story"

Literary Elements

In this chapter, we'll look at these elements and describe both their own nature and how they relate to one another. We'll "peel back" the surface of several works of fiction and examine their basic working parts to demonstrate how the elements work to improve the story.

To begin, we'll more closely explore a concept introduced in Chapter Two: the idea of **theme**. Theme comes from the successful working of all the different parts and stays in our memories long after other details fade.

THEME

We have already discussed how theme is the "big idea" an author wants to share with his audience, a work's central message communicated by the entire story. Themes most often offer a comment or idea about human nature, life, and society.

We may criticize a too obvious message. For example, have you ever watched a movie or television program and thought, "That was so obvious" When a work's theme is too predictable, we grow impatient because our mind feels "cheated" reaching its own conclusions.

When a theme is obviously meant to reach an audience, we say it is an **explicit** theme. Fables and folk tales often include explicit or directly stated themes. The theme of "The Tortoise and the Hare" is clearly understood to be "slow and steady wins the race."

Themes may also be **abstract**, which in this sense means open to reader interpretation. Abstract themes usually appear in **genre** fiction such as science fiction or detective novels. But remember, any work of *literature* — which is to say, a story of great fame and stature — has a theme. For example, the implied theme of *Charlotte's Web* is that loyalty and friendship conquer obstacles too big for one person to face alone.

Sometimes great works have many themes, and scholars argue about the author's purpose in creating them. After 150 years, people are still discussing the themes of *Moby Dick*.

A theme is something the mind discovers for itself after considering the story as a whole. Do not confuse a book's theme with its subject. The **subject** is *who* and *what* the story is about — the characters places, and events.

FINDING THE THEME IN FICTION

A **theme** is a thread of feeling or belief that runs across the entire work, pulling all the events and objects of the subject into a whole. Sometimes this message is difficult to discover, like a message hidden in a bottle. That is when the reader needs to put together all the details in the work. Once you have assembled the story's many details and ideas, the theme will often present itself clearly and with focus, rising up from the story into your mind for you to consider.

When asked a question about theme, you may be tempted to answer by describing what happens in the story. But don't mistake *plot* for the *message* that the author conveys through the story to his audience. As an example, let's consider literature's most famous love story:

The Tragedy of Romeo and Juliet	
Subject:	**Theme:**
Two young people fall in love, despite their families' wishes.	Love conquers everything, even the world around us.

On the left, you see the events of the story itself. But the theme (the "message") applies not just to the story but to life in general — at least, according to the author. That message is the **theme**.

Some situations are shared by all peoples no matter when and where they live. Literature often focuses on these **universal themes** of common human understanding. For a theme to be universal, it must deal with human experiences that are found in any particular time period or cultural environment.

Some of the most common universal themes include good struggling against evil, the delicateness of life, and, of course, love in all of its varieties. Family relationships, death and rebirth, man surviving against nature, and a young person's struggle to reach maturity (in works known as "coming of age" stories) are also common universal themes.

Universal Themes in Fiction	
Theme	**Work**
Good always wins over evil	*The Chronicles of Narnia*
Power corrupts	*Animal Farm*
War forces boys to become men	*Johnny Tremain*
Nature works by its own set of laws	*The Old Man & The Sea*
Individuals must think for themselves	*Ender's Game*

Literary Elements

Practice 1: Identifying Universal Themes

Make a list of three novels or short stories you have read and three movies you have seen. Next to the title of each one, identify the universal theme. Write a sentence or two explaining why you decided on your answer. Compare lists with your classmates. If you listed some of the same titles but have different answers, discuss and defend your answers.

SETTING

Setting is the background for the action of a story. Setting includes the *time*, *place* and *general surroundings* in which the story takes place. A setting can be realistic, as it would be in a historical novel, or it can be imaginary, as in science fiction or fantasy. For example, the New York City found in an episode of *Law & Order* would be for the most part true to life, while the world of *Shrek* is purely make-believe.

The setting of a story affects the *mood*, creates *conflict*, and influences the *characters*. Below are three aspects of setting.

Time:	when the story takes place. It may be past, present, or future. For example, the novel *Ender's Game* takes place at least a hundred years in the future.
Place:	where the story happens, including such details as geographic place, scenery, or arrangement of a house or room. The place may be real or imaginary. In S.E. Hinton's *The Outsiders*, the action takes place in Tulsa, Oklahoma during the 1950s.
General Surroundings:	the daily habits of characters, including their job, religious practices, or the economic or emotional spirit of the area in which they live. In *Ender's Game*, the earth has twice been attacked by aliens, and the nations of the world work together to prevent a third invasion. These conditions affect the main character's life in every possible way.

Practice 2: Setting

Read the passage below carefully. Then answer the questions which follow it.

> Paris was blockaded, starved, in its death agony. Sparrows were becoming scarcer and scarcer on the rooftops and the sewers were being depopulated. One ate whatever one could get.
>
> As he was strolling sadly along the outer boulevard one bright January morning, his hands in his trousers pockets and his stomach empty, M. Morissot, watchmaker by trade but local militiaman for the time being, stopped short before a fellow militiaman whom he recognized as a friend. It was M. Sauvage, a riverside acquaintance.
>
> – excerpted from "Two Friends" by Guy de Maupassant

1. Short answer. Use your own paper to respond. The story takes place during a war between Germany and France in 1870. Write a list of words and phrases in the description of the setting, which indicate that the story takes place during a war.

2. Short answer. Use your own paper to respond. What details from the passage describe the characters and their lives?

CHARACTERIZATION

In literature, characters must have clear qualities which set them apart from characters in the same work or characters in other works. Authors must present these qualities in such a way that shows what to expect of the characters' behavior.

Characterization consists of the statements an author makes about a character through description or narration. It also includes what can be observed about a character, such as how the character speaks, other characters' opinions of him, his actions, and how he reacts to others.

Revealing Character Traits	
Description	How characters look, dress, and what their ages are, just as you might describe a friend of yours to someone. In Eudora Welty's story "A Worn Path," the narrator describes the main character, Phoenix Jackson, as an old, small Negro woman in plain but neat clothing. This reveals both her poverty and her self-respect.
Narration	The telling of the story through a speaker. The speaker could be one of the characters or could be an unknown observer. The speaker will tell how other characters feel or think about another character or will describe how they act towards that character. In *The Red Badge of Courage*, there is an unknown narrator who is limited to telling the story through the eyes of a young soldier.
Dialogue	Conversation between two or more people. People in literature speak to each other as people in your class do. Mark Twain in *Huckleberry Finn* shows the character traits of Huck and Jim in the talks they share while they float down the Mississippi River.
Actions	Sometimes the actions of a character speak louder than words to show the character's true self. The same is true of politicians and your best friends. The main characters in O. Henry's "The Gift of the Magi" show their love for each other by placing the happiness of the other before themselves.

Different characters also play different roles over the course of a story:

- The **narrator** is the person telling the story. He or she will often be the main character. In *The Outsiders*, the narrator is Pony Boy Curtis, and he is also the main character. Sometimes another character or an outside voice narrates the story, as for example Watson in *The Hound of the Baskervilles*.

- The **protagonist** is the main character. He leads the plot, gets involved in the conflict and often changes by the end of the story. The protagonist is usually the hero but not always. Many short stories by Edgar Allan Poe and novels by Stephen King have protagonists who do evil things.

- The **antagonist** struggles against the protagonist. An antagonist can also be a force blocking the protagonist, such as nature or society. In Jack London's "To Build a Fire," the antagonist is the freezing climate of the Alaskan wilderness.

Literary Elements

In addition to these central "actors," a novel may also feature many other characters. A short story will have only a few, but they all act according to the same kinds of influences.

Influences on Characters	
Relationships	The character's background and contact with other people. Through the narrator of a story, authors can describe a character's family life, job, and social position. The author can show through actions and words how a character thinks and feels about the story's relationships (family, friends, strangers). Often, these elements are also part of the story's setting. Tom Sawyer's friendships with Huck Finn and Becky Thatcher shape much of his young life.
Motivations	The reasons that characters have for acting a certain way, often wants or desires. In human nature, we often find that the desire to *keep* something is often every bit as powerful as the desire to *obtain* something. For example, a character who is enslaved will want to gain freedom. A character who has freedom will want to keep it. In John Steinbeck's *The Pearl*, Kino and his family want to get out of poverty.
Conflicts	The obstacles or problems that the characters must resolve. How characters deal with such problems define them as people. Do they run from difficulty, or do they rise above? *Internal conflicts* are created inside the character's mind when trying to decide on the right way to act or how to understand life. *External conflicts* come from outside the character, like an upcoming test, or a street fight with a bitter rival. Protagonists usually struggle with both kinds of conflict, and a good story will play off the tension between the two. In *The Old Man and the Sea*, the fisherman Santiago struggles to catch a giant marlin in the Gulf of Mexico while at the same time fighting feelings of doubt and hopelessness.

As characters act, they probably behave according to one of a set of character **types**, depending on how they behave over the course of a narrative.

Types of Characters	
dynamic	a character that undergoes a significant change or personal growth during the time frame of a literary work; a dynamic character can also be called a round character.
static	a character that does not experience change or growth during the time frame of a literary work; at the end of a work, they are essentially the same as at the beginning of the work; they are also called a flat character.
round	a multi-dimensional character; reader knows many personality aspects of the character; usually a dynamic and/or a major character
flat	a one-dimensional character; reader has limited knowledge about the character; often a stereotype and/or a minor character and/or static. Flat characters are often background people on the street, in stores, etc.

Chapter 4

Practice 3: Characters and Characterization

Read the following passage, and answer the questions that follow.

The transcontinental express swung along the windings of the Sand River Valley, and in the rear seat of the observation car a young man sat greatly at his ease, not in the least discomfited by the fierce sunlight which beat in upon his brown face and neck and strong back. There was a look of relaxation and of great passivity about his broad shoulders, which seemed almost too heavy until he stood up and squared them. He wore a pale flannel shirt and a blue silk necktie with loose ends. His trousers were wide and belted at the waist, and his short sack coat hung open. His heavy shoes had seen good service. His reddish-brown hair, like his clothes, had a foreign cut. He had deep-set, dark blue eyes under heavy reddish eyebrows. His face was kept clean only by close shaving, and even the sharpest razor left a glint of yellow in the smooth brown of his skin. His teeth and the palms of his hands were very white. His head, which looked hard and stubborn, lay indolently in the green cushion of the wicker chair, and as he looked out at the ripe summer country a teasing, not unkindly smile played over his lips. Once, as he basked thus comfortably, a quick light flashed in his eyes, curiously dilating the pupils, and his mouth became a hard, straight line, gradually relaxing into its former smile of rather kindly mockery. He told himself, apparently, that there was no point in getting excited; and he seemed a master hand at taking his ease when he could. Neither the sharp whistle of the locomotive nor the brakeman's call disturbed him. It was not until after the train had stopped that he rose, put on a Panama hat, took from the rack a small valise and a flute case, and stepped deliberately to the station platform. The baggage was already unloaded, and the stranger presented a check for a battered sole-leather steamer trunk.

– excerpted from "The Bohemian Girl," by Willa Cather

1. Which of the following sentences best describes the young man?

    A. He is poorly dressed and feeling uncomfortable.
    B. He is hard working and curious.
    C. He is dressed strangely and very excited.
    D. He is strong, relaxed, and a stranger in this place.

2. What does the following sentence say about the character?

    "His head, which looked hard and stubborn, lay indolently in the green cushion of the wicker chair, and as he looked out at the ripe summer country a teasing, not unkindly smile played over his lips."

    A. He is tired from his long trip and wishes it were over.
    B. He is a stubborn man and is feeling pleased.
    C. He enjoys teasing people.
    D. His head is very large and strange looking.

3. What three methods of characterization are used to describe the man?
    A. observation, dialogue, and narration
    B. dialogue, description, and action
    C. description, observation, and action
    D. thought, observation, and narration

4. On your own paper, find and write an example for each of the three methods of characterization used in this passage.

Literary Elements

Practice 4: Characters and Characterization

Choose a fictional work from your literature textbook. After reading your selection, complete the following activities on your own paper.

1. Describe the character types from the fictional work you selected. Who is the narrator? Who is the protagonist? Who or what is the antagonist?

2. Choose one main character from your fictional work, and describe how the author has shown the character. Which of the four methods of characterization did the author use to reveal the character? Then, describe the character completely by listing as many character traits as you can.

POINT OF VIEW

Point of View is the perspective, or outlook, from which a writer tells a story. There may be a character narrating the story, or there may be an unknown, all-seeing speaker describing the action and thoughts of the main characters. It may help you to think of point of view as a "camera" with which the author shows you the story. The views shown by the camera are usually one of three perspectives:

	FEATURES OF POINT OF VIEW
First Person	The narrator tells the story from his own point of view, saying "I did this" or "I did that." Perhaps the most famous example of recent times is J.D. Salinger's *The Catcher In The Rye*.
Second Person	The book itself addresses the reader, as if the reader is an active character in the book. For example, "You are walking down the street one morning when…" Second person narration is rarely used. Jay McInerney's *Bright Lights, Big City* is one of only a few examples of second person narration.
Third Person	This point of view contains the majority of fiction written before the 20th century. In third person, a narrator moves unseen among the characters, relating their actions. There are two kinds of third person narration, with different advantages and difficulties:
• omniscient	narrators can see everything and everywhere, even relating the characters' thoughts. Charles Dickens' *Oliver Twist* is an example of omniscient third.
• limited	third person narration (sometimes called **approximate third**) centers on one character and observes only what he sees, hears, feels or does. It will also sometimes include his thoughts. *The Red Badge of Courage* is an example of this kind of work.

Practice 5: Point of View

Go through your library, finding two examples of each type of point of view. Read a few paragraphs of each, listening to the "feel" of the narration. Which one feels most comfortable to you? Why? On a separate piece of paper, record your thoughts. Then share them with a teacher or classmate.

Tone In Fiction

As discussed in Chapter One, **tone** is the attitude an author takes towards his subject. In fiction, tone can run the full range of human emotion, though in most works the author adapts a tone of at least mild sympathy towards the characters.

For example, in the novels of John Steinbeck (*The Grapes of Wrath, The Pearl, Of Mice and Men)*, the characters are always at the mercy of larger forces upon which they have no control. Steinbeck has great compassion for the people caught in the harsh realities of every day life, and expresses them through deep caring and sympathy for the men and women who populate his work.

Mood is similar to tone, in that it is the atmosphere of the world in which the characters inhabit. Like tone, there is no limit to the range of mood. Some prime examples of excellent mood in books include the novels of Raymond Chandler, the short stories of Rick Bass, and the memoirs of David Sedaris. Poe's short stories also convey a somber mood.

Practice 6: Tone

Go through your literature text, finding five short stories or excerpts you enjoyed. Scan through them again, taking notice of the authors' tones and the stories' moods. Write the mood and tone next to the title on a separate sheet of paper. Which tones and moods did you enjoy more than others and why?

Plot

Perhaps more than any other element, **plot** is essential to storytelling. It is the pattern of events in a story, everything that happens to form a narrative. It is important to distinguish between plot and story, however. A story is the events that happen, one building on another; a plot is the events *plus* the characters and their reactions, the setting, and whatever changes happen to either. Consider this example to tell the two apart:

"The king died, and then the queen died" is a *story*.
"The king died, and then the queen died of a broken heart" is a *plot*.

Traditionally, a plot has several parts. There is the **introduction**, the **rising action**, the **conflict**, the **climax**, the **falling action**, and the **resolution**.

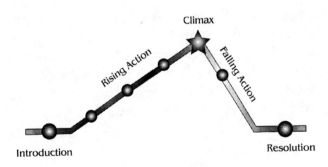

Literary Elements

Introduction   often called *exposition*, the introduction is the opening of a story. The author describes the setting, introduces the characters, and reveals conflict. Some authors choose to introduce a main character after revealing the conflict. For example, F. Scott Fitzgerald's short story "May Day" begins with an elaborate description of New York, then "zooms in" to several different groups of characters and their conflicts.

Conflict   occurs when the character encounters an obstacle to something he needs or wants. The struggle that takes place may be between a character and nature, between a character and himself (inner conflict), a character and other characters, or between the character and society — its laws, or its expectations and pressures.

Four Types of Conflict

Conflict Type	Novel/Short Story
Man vs. Man	*The Sea Wolf*
Man vs. Society	*1984*
Man vs. Nature	*Lord of the Flies*
Man vs. Self	*The Miracle Worker*

Rising action   is the increasing tension and pressure the character feels to get the object of his need or desire. The rising action of *Romeo and Juliet* occurs when the two lovers attempt to be together despite their families' feud. When rising action reaches its highest degree of suspense, it reaches the *climax*.

Climax   is the turning point in a story. It may occur when the conflict is at its worst or when circumstances permanently change for the character. For example, at the climax of John Steinbeck's *The Grapes of Wrath*, the protagonist Tom Joad vows to spend his life avenging the suffering of poor people. His fate is sealed and his life changed forever afterwards.

Falling action   is the easing of pressure on the main characters: the gradual return to a normal (possibly better) pattern in their lives. In *The Return of the King*, the falling action includes Aragorn becoming king and Frodo returning to the Shire. The falling action is almost always a shorter segment of the story than the rising action.

Resolution   also sometimes called the *denouement*, resolution is the very ending of a story. It is the point at which all conflict has been settled. In tragedy, the resolution is often a sad but final one. In Shakespeare's *MacBeth*, most of the main characters are killed by the final scene. However, in a comedy, the resolution is usually a happy solution to all problems. This famous "happy ending" in many comedies involves a wedding, an event which by its very nature promises a new life for the main characters.

SUBPLOTS

In almost all novels and most short stories, at least one smaller story will take place as the main plot unfolds. These **subplots** are not necessary to the main story but reinforce it, so that the reader's interest holds and deepens. In many stories, the subplot takes the form of a love story between the principal male and female characters. A subplot may also tell how two rivals learn to work together for a greater good. Subplots can include their own climax, one that dovetails into the climax of the main plot. Some subplots you may recognize include:

Subplots	
Book:	**Subplot:**
Harry Potter and the Goblet of Fire	Hermione's quest to help the house elves
Lord of the Flies	The search for the Beastie
Tom Sawyer	Tom's romance with Becky Thatcher
Flowers for Algernon	Charlie getting along with his co-workers

Practice 7: Conflict

Carefully read the passage, observing the type of conflict, and answer the questions below.

"Dread" Locks

Elizabeth sat on the stool in the center of the room, terrified of moving. The scissors seemed to be on all sides of her.

She must be patient, she kept telling herself. No moving. Be good. One small move and she might lose an ear, or an eye. How much longer was this going to take? She realized she had been holding her breath and let a sigh escape.

"Be still," the voice said.

"Are you almost done?" she dared to ask.

"Yes, we're almost done. Keep still, and it will go faster."

Time seemed to slow down. She saw the hair fall to the ground and began to worry. Elizabeth looked out the window and let her mind wander. She thought of sunshine, swings, slides, and sodas.

"Okay, you're all done Elizabeth. What do you think?" asked the woman handing her a mirror.

Elizabeth looked in the mirror and slowly a smile spread across her face. She looked okay. The same girl still looked back at her from the mirror.

Literary Elements

"You look great, Lizzie. Now let's get your treat and head to the park. Thank you for being so good," said her mother, taking her hand and helping her off the stool.

1. The main conflict in this passage is between Elizabeth and

    A. her mother.   B. herself.   C. society.   D. nature.

2. **Short Answer.** Use your own paper to write your response. Explain the conflict from the passage. What is Elizabeth struggling against? Use details from the passage to support your answer.

Practice 8: More Conflict

On your own or with your class, complete the following activities. Review the answers with your teacher.

1. Recall some of the characters and scenes from *The Wizard of Oz*. Describe the plots and subplots in the film.

2. Discuss one or two recent popular films. Describe the main plot and any subplots.

3. Choose a short story or novel you recently read in school. Identify the plot and subplots in these works. Or find them in one of the following novels or stories:

Novels	Short Stories
Across Five Aprils	"Masque of the Red Death"
Flowers for Algernon	"The Lottery"
The Red Badge of Courage	"Monkey's Paw"
The Pearl	"The Rocking Horse Winner"

CHAPTER 4 SUMMARY

Fiction has working elements that have not changed for centuries. They include theme, setting, characterization, point of view, tone, and plot. These moving parts are found in all literary genres and are central to the storytelling process. When reading fiction or watching movies and television, close attention to these elements will reveal the author's goals and skills.

Theme is the message of the story, the idea an author wants to share with his audience. If a theme is *too* obvious, the audience may grow impatient. Most works of literature have at least one theme. Some themes are **explicit** (deliberate) and others **abstract** (open to reader interpretation) Some themes appear everywhere, in all cultures, and are considered **universal**.

Setting is the background of the story's action. It includes the **time**, **place**, and **general surroundings** in which the characters live and operate. A setting can be real or imaginary.

Characterization is the method an author uses to get us to care about the characters in a story. It includes what can be observed and what can be inferred about anyone in a story. Character roles include **protagonist**, **antagonist**, and **narrator**. Several forces drive a character, including their relationships with others, their conflicts, and their background. A character is revealed to us through his or her description, dialogue with others, and action. What a character does determines his type: dynamic, static, round or flat.

Point of view is the writer's perspective in the story.

Tone involves the writer's attitude towards the subject material.

Plot is the events that happen along the course of a story. Plot develops through conflict, with rising tension leading to a climax in which the conflict is resolved. Plot is not the same as story, which is all the elements of setting, mood, conflict and character together. Rather, it is a "**timeline**" of events as they happen.

- A **subplot** is a smaller story that takes place while the main plot develops. It is sometimes a love story involving two of the characters.

Literary Elements

CHAPTER 4 REVIEW

Read the following passages. Then choose the best statement of the theme.

One day a blind man was walking along a road. Because he could not see, he was feeling his way with a stick. He came to a muddy place in the road; and he could not find a way to get across it. The blind man sat down by the side of the road.

Pretty soon a lame man came very slowly down the road. The blind man heard him and spoke.

"Please help me. I am blind and I can't get through this mud."

"I wish I could help you, but I am lame and I can barely walk myself," said the lame man as he sat down beside the blind man to rest.

"You look like a very strong man," said the lame man to the blind man.

"I am very strong," said the blind man, "but it is difficult for me to find my way because I cannot see."

"I have good eyes, but it is difficult for me to get around because I am lame."

The lame man had an idea. He said to the blind man, "Let us help each other. You carry me with your strong body. I will be your eyes and tell you where to go."

The blind man helped the lame man get up on his shoulders. They went past the mud and down the road very happily helping one another.

– based on an Aesop fable

1.
- A. Never let the blind lead the blind, nor the lame lead the lame.
- B. The world is a much happier place when people help each other.
- C. One person can make a difference in the world.
- D. Work is quicker with more hands.

The mother and child walked hand in hand, enjoying each other's company. They stopped occasionally to examine one thing or another: a spider in its web, a flower in full bloom, a caterpillar on a leaf. As they approached an older woman sitting on a bench in the shade of a tree, they stopped for a rest. The woman smiled at them. After a brief rest, the mother gathered their things and prepared to leave.

"Treasure these times; they pass so quickly," the old woman said with kindness.

"I will. I do," replied the mother. She collected her child and continued on their walk. She made sure to hold her child's hand and slow her pace to match the little feet. They walked on with the sun upon their faces and the world all around them.

2.
- A. Wisdom comes with age, and elders should be respected.
- B. Exercise is important.
- C. The small things in life should be appreciated for memory's sake.
- D. Old people are kind.

Chapter 4

3. **Short Answer.** Pick three short stories, books or poems you have read in class this year. On a separate sheet of paper, write down any explicit themes in one column. Then, in two more columns, see if you can identify any abstract themes. Do the works you've selected relate any universal themes to the audience? If so, how?

Setting. Read the following passage. Then, answer the questions that follow.

During the whole of a dull, dark, and soundless day in the autumn of the year, when the clouds hung oppressively low in the heavens, I had been passing alone, on horseback, through a singularly dreary tract of country; and at length found myself, as the shades of the evening drew on, within view of the melancholy House of Usher. I know not how it was — but, with the first glimpse of the building, a sense of insufferable gloom pervaded my spirit. I say insufferable; for the feeling was unrelieved by any of that half-pleasurable, because poetic, sentiment, with which the mind usually receives even the sternest natural images of the desolate or terrible. I looked upon the scene before me — upon the mere house, and the simple landscape features of the domain — upon the bleak walls — upon the vacant eye-like windows — upon a few rank sedges — and upon a few white trunks of decayed trees — with an utter depression of soul...

– excerpted from "The Fall of the House of Usher," by Edgar Allan Poe

4. Which of the following best describes the setting?
    A. a quiet, rainy day in the country
    B. a sunny afternoon at the Usher House
    C. a dark, gloomy day in autumn at the House of Usher in the country
    D. a simple landscape of trees, hedges, and quiet buildings

5. **Short answer.** Use your own paper for your response. What setting details from the passage show the mood? Do the ideas work together to give a vivid sense of the location? Do you think they imply what will happen in the story? Why or why not?

Characterization. Read the following passage. Then answer the questions that follow.

I (heard) the story, bit by bit, from various people, and, as generally happens in such cases, each time it was a different story. If you know Starkfield, Massachusetts, you know the post-office. If you know the post-office you must have seen Ethan Frome drive up to it, drop the reins on his hollow-backed bay and drag himself across the brick pavement to the white colonnade: and you must have asked who he was.

Literary Elements

It was there that, several years ago, I saw him for the first time; and the sight pulled me up sharp. Even then he was the most striking figure in Starkfield, though he was but the ruin of a man. It was not so much his great height that marked him, for the "natives" were easily singled out by their lank longitude from the stockier foreign breed: it was the careless powerful look he had, in spite of a lameness checking each step like the jerk of a chain. There was something bleak and unapproachable in his face, and he was so stiffened and grizzled that I took him for an old man and was surprised to hear that he was not more than fifty-two. I (heard) this from Harmon Gow, who had driven the stage from Bettsbridge to Starkfield in pre-trolley days and knew the chronicle of all the families on his line.

"He's looked that way ever since he had his smash-up; and that's twenty-four years ago come next February," Harmon threw out between reminiscent pauses.

— excerpted from *Ethan Frome*, by Edith Wharton

6. How do you learn you about Ethan Frome? Select all that apply.
   A. his actions
   B. his thoughts
   C. his appearance
   D. the comments of others

7. Which is a good description of Ethan Frome?
   A. a strong, powerful man
   B. an old, weak man
   C. a tall, lame man
   D. a strong, lean young man

8. **Extended Response.** Use your own paper to write your response. Write a detailed character description of Ethan Frome. Use details from the selection to support your description. What does Ethan Frome's description tell us about the man?

Tone. Select the correct answer from the choices given.

9. Look back on the selection from "The Fall of the House of Usher." Which statement best describes the mood created by the language?
   A. thoughtful and lonely
   B. content and tranquil
   C. cheerful and excited
   D. dark and foreboding

10. Think of five works of fiction you have read or watched in the past year. On a separate sheet of paper, write down their moods. Now, write the opposite mood in a third column on your page. Consider how the works would be different if the moods were changed. Discuss your results with friends or classmates.

Plot. Read the following passage. Then answer the questions that follow.

It was a dark autumn night. The old banker was walking up and down his study and remembering how, fifteen years before, he had given a party one autumn evening. There had been many clever men there, and there had been interesting conversations. Among other things they had talked of capital punishment. The majority of the guests, among whom were many journalists and intellectual men, disapproved of the death penalty. They considered that form of punishment out of date, immoral, and unsuitable for Christian nations. In the opinion of some of them the death penalty ought to be replaced everywhere by imprisonment for life.

"I don't agree with you," said their host the banker. "I have not tried either the death penalty or imprisonment for life, but if one may judge — a priori*— the death penalty is more moral and more humane than imprisonment for life. Capital punishment kills a man at once, but lifelong imprisonment kills him slowly. Which executioner is the more humane, he who kills you in a few minutes or he who drags the life out of you in the course of many years?"

"Both are equally immoral," observed one of the guests, "for they both have the same object — to take away life. The State is not God. It has not the right to take away what it cannot restore when it wants to."

Among the guests was a young lawyer, a young man of five-and-twenty. When he was asked his opinion, he said: "The death sentence and the life sentence are equally immoral, but if I had to choose between the death penalty and imprisonment for life, I would certainly choose the second. To live anyhow is better than not at all."

A lively discussion arose. The banker, who was younger and more nervous in those days, was suddenly carried away by excitement; he struck the table with his fist and shouted at the young man:

"It's not true! I'll bet you two millions you wouldn't stay in solitary confinement for five years."

"If you mean that in earnest," said the young man, "I'll take the bet, but I would stay not five but fifteen years."

"Fifteen? Done!" cried the banker. "Gentlemen, I stake two millions!"

"Agreed! You stake your millions and I stake my freedom!" said the young man.

And this wild, senseless bet was carried out! The banker, spoilt and frivolous, with millions beyond his reckoning, was delighted at the bet. At supper he made fun of the young man, and said:

"Think better of it, young man, while there is still time. To me two millions are a trifle, but you are losing three or four of the best years of your life. I say three or four, because you won't stay longer. Don't forget either, you unhappy man, that voluntary confinement is a great deal harder to bear than compulsory. The thought that you have the right to step out in liberty at any moment will poison your whole existence in prison. I am sorry for you."

– excerpted from "The Bet," by Anton Chekhov

* without previous experience

11. Which point of view does the author use in this selection?
    A. third person limited
    B. third person omniscient
    C. second person
    D. first person

Literary Elements

12. Which of the following is the *best* statement of plot for this passage?
    A. A banker remembers when he bet a young lawyer that the young man could not stay imprisoned for 15 years.
    B. A banker argues with a young lawyer about capital punishment and bets him that life imprisonment is worse than the death penalty.
    C. A banker remembers a party he held 15 years ago. His guests approved of the death penalty, especially a young lawyer. They had a lively discussion.
    D. A banker is delighted to bet a lawyer he can't stay imprisoned for 25 years.

13. What is the *best* statement of the conflict present in this passage?
    A. The guests are irritated by the discussion at the party.
    B. A banker and a young lawyer argue about death vs. life sentences
    C. Which is the better choice: capital punishment or life imprisonment?
    D. The banker is bothered by his party guests' opinions.

14. Which of the following *best* describes the setting of the action in this passage?
    A. a dark, autumn night in a banker's study
    B. an argument at a party
    C. a wealthy banker's home one evening
    D. a party one autumn evening at a banker's home, 15 years ago

15. Which statement of theme might be hinted at in the last paragraph?
    A. Imprisonment poisons a person.
    B. It is impossible to survive confinement.
    C. The prisons we create for ourselves are hardest to bear.
    D. The best years of our lives should not be squandered by gambling.

16. **Extended Response.** Think of your favorite story of the literature you have studied this year. Was the plot something you found easy to predict? Why or why not? On a separate sheet of paper, write a timeline, putting the introduction on one end and the resolution on the other. In between, write down what you felt were the major events between the two. Share your answers with a friend or classmate.

Web Sites

http://gutenberg.net/index.html
Project Gutenberg is the Internet's oldest producer of free electronic books, of both fiction and non-fiction. This is an index of the complete works that you can find on the site. You may search by author, title, or book. The index is in a "Winzip" file, so there may be certain programs which can deal with the index better than others.

http://etext.lib.virginia.edu/ebooks/ebooklist.html
The University of Virginia's free online ebook library has links to thousands of complete texts that readers can view with Microsoft. There are also hundreds of other texts that can be read with a palm device. There are some journals linked to the site, such as a journal of essays about history. The site is fairly easy to navigate and has illustrations to break up the text.

http://www.umich.edu/~umfandsf/symbolismproject/symbolism.html/index.html
A dictionary of symbolism from the University of Michigan, this huge site includes meanings for all kinds of symbols and themes in literature. The symbols are arranged in alphabetical listings for easy access. The site also has pages about monsters and visual symbolism.

Literary Elements

Chapter 5
Consumer, Public, and Workplace Documents

This chapter covers the following content standards:

1.C.3	Students will analyze and draw accurate conclusions about information contained in warranties, contracts, job descriptions, technical descriptions, and other information sources, selected from labels, warning manuals, directions, applications and forms in order to complete specific tasks.

Have you ever picked up a product catalog or brochure from the local computer store and found the text overwhelming? The information presented is not simple: such highly technical reading is known as **informational materials** and are generally designed to inform and instruct. In the 21st century, they are becoming more commonplace every day.

This chapter will introduce you to informational materials, including those that contain **graphics** and **visual aids**. These types of publications, which come from private, public, and government sources, must be read and understood for any number of reasons: filing taxes, planning trips, or getting jobs, to name just a few.

You will also be shown how to analyze each document for the specific purpose it serves and how to **synthesize** and draw conclusions from these sources. The term *synthesize* means to bring different parts together to create something new and complete. In reading, synthesizing information aids in understanding various texts and documents.

INFORMATIONAL MATERIALS

These are materials you'll almost certainly need in your lifetime and are largely issued or handled by government and private **bureaucracies** (administrative systems with rigid and complex procedures). Preparing for their application processes in advance will save you time, money, and untold amounts of stress.

Consumer, Public, and Workplace Documents

Consumer documents are forms and materials used in the purchase, maintenance, and conditions for use of products or services.

Contracts

are legal documents that hold buyers and sellers to agreements or promises. Remember that contracts are usually very complex and often intentionally vague with text written in dense, legal language. Contracts require all vital personal information: name, address, phone number, date of birth.

Product Information Guides

give valuable facts about a certain product or type of goods. These may include definitions, descriptions, features, various uses, technical support information, order forms, and cautions or warnings. Such guides are available online through private companies and from the government.

Instruction Manuals

are general how-to guides or instructions on an activity or process. Manuals can be any size from one-page brochures to large books. Consumer instruction manuals give exact directions on how to install and/or operate a product. Manuals may have tips on how to fix problems or may simply list where to call for help. They often contain diagrams and pictures.

Warranties

are very exact, legally binding guarantees about the level of performance a product should display. They are also promises about what kind of service a consumer can expect if a product has problems or defects. Warranties are almost always packaged with products and are included in product information.

Chapter 5

Personal documents include both manuals and information-gathering forms. Most national and state governments expect workers to complete forms supplying personal and sometimes legal identity information. This information includes money matters, business matters, immigration matters, and even family data (births, deaths, and so on).

Tax Forms **are a record of how much money you have made in a certain year and how much of that money you need to pay the government.** The simplest of the tax forms is the 1040 EZ. All taxpayers fill out a W-4 form each year. Your employer gives you the W-4 to choose how much tax you want taken out of your paycheck; you pay the remaining tax by April 15th or receive a refund if you paid too much. There are dozens of other tax forms for different situations.

Identity Forms The most important identity form, besides a driver's license or ID card, is a **Social Security Number**. These federally-issued identification numbers are required to work in the United States. In most cases, parents fill out registration forms for a SSN along with birth certificates. New citizens need to fill out their own Social Security forms, and if people change their names (like when a woman gets married), the forms must be changed. **Passports** and **immigration numbers** are also important. All the forms to apply for these numbers ask for personal information and citizenship status.

Licenses A **driver's license** is necessary for jobs such as delivery and trucking and mandatory for anyone operating a vehicle on public roads. Nurses, teachers, electricians and people in other highly-skilled professions must also pass an exam to obtain their **professional licenses**.

Insurance Forms enable you to get health, life, and dental insurance when you obtain a full-time job. These forms will ask for your medical history and the usual personal information. There are also reports to fill out in case an employee is injured on the job called **workers' compensation** and are part of the employer's insurance plan.

Consumer, Public, and Workplace Documents

Employee Notices

are a means for companies to make announcements to employees. These notices may announce changes in **company policy** (such as uniforms or name badges), or **educational opportunities** (such as Spanish lessons at a local college). Increasingly, employee notices are sent out via email. Some notices require a response, but many do not.

Work Manuals

When a company has many specific procedures and rules, a **workplace manual** may be issued to all employees. There will be a table of contents, graphics, and written directions to teach employees the detailed requirements of the company. For example, your state highway department issues manuals on the proper operation of construction vehicles such as tractors, bulldozers, and dump trucks.

Public documents are forms and documents available to help citizens in everyday business and errands. Many of these are reference materials, such as phone books and informative pamphlets.

Bank Documents

Many banks have **savings and checking accounts** for people to choose from. All accounts require that you fill out forms. These forms ask for the usual personal information such as address, date of birth, and SSN. After setting up an account, you will need to fill out deposit slips, and you should balance your bank book or check book often. Banks can also provide materials for college savings plans through different financial groups.

Loan Documents

You may already be thinking of how to pay for college or buy a car. Some people sign loan agreements to help with these expenses. In school, your counselors can help with student loan applications. Be careful of other loans, though, such as for cars and credit cards. Loan documents are financial contracts and are legally binding. Also, credit card companies are targeting younger consumers who may not realize the workings of interests rates and payment deadlines, so be wary of free gifts or prizes for filling out credit card applications.

Communication and Travel

Cell phone companies, Internet service providers, and long distance companies all have instructional or informational texts to read. Telephone use in particular has become more complicated with many different plans for cell phones, family plans, overages, and monthly minute allowances.

Public transportation such as trains, planes, and buses are a complicated part of everyday life. Being able to read and understand schedules, ticket information, or lease agreements is an important skill. As with contracts, be careful to thoroughly understand any service agreement before signing it, and ask someone to explain schedules that appear confusing. It's not easy to ask for help — and you might feel less than intelligent — but you'll feel worse later on, if you've missed opportunities or spent more money than necessary.

ANALYSIS OF WORKPLACE DOCUMENTS

As you can see, there is a form or document for almost everything, especially in the workplace. Imagine that you are applying for a summer job at a citywide delivery firm. First, you must fill out the job application. If you are hired and are going to be driving, you must get a commercial driver's license. You must also fill out forms for insurance, for income tax, and possibly for uniform allowances. The company will most likely issue you an employee manual. This manual will have instructions on how to fill out insurance claim forms, work reports, make requests for time off, and sign up for automatic payroll deposits.

STRUCTURE

The **structure** of a document refers to the way its parts are arranged. Most documents have sections for personal information, for explanations of the document, and for signatures. How these parts or sections are arranged can help you to understand the purpose and intent of the document's author. The leave form at right is **informal** in structure and information.

- The most important information is the identity of the employee: your name.
- There are two lines to identify your area within the company.
- Next are the lines for the requested leave.
- The signatures make the request official.
- The other important information, the approval or disapproval, comes last. Boxes are used to focus on this information.

Nonamé Delivery Co.
Vacation or Sick Leave Permission Form
Date_____

Name _____
Section ID # _____
Supervisor _____

Beginning Date of Leave _____
Ending Date of Leave _____

Employee Signature _____
Supervisor's Signature _____

Approved ☐   Disapproved ☐

Comments: _____

Form 1

Consumer, Public, and Workplace Documents

The **Employee's Withholding Allowance Certificate**, known informally as the **W-4**, serves a very different purpose from the permission form's purpose. It is very **formal** in structure and purpose.

Form 2

- This is a federal government form. Its identification is entered first, reading left to right, by form number and federal office.

- Notice the year is entered along the first line in very bold type. Both the position of the type and the bolded, large size of the type makes it very clear that this is an important item on the form.

- The different sections in the body of the form are important enough to be numbered within the form. Section **1** is for personal information. Section **2** asks for your social security number, section **3** asks for your marital status, and section **4** asks for details about your social security card. All these items are part of your legal identity.

- In the middle of the form are the sections **5 – 7**. These are the sections concerning the details of your wages and the amount of tax you want withheld from your paycheck. The boxes and arrows help bring notice to this important part of the form.

- There is a large signature box with warnings about how important the signature is for the legal purpose of the form. The last section is for your company's identity, including an identification number.

The **W-4** is a very formal document, with legal as well as personal identity required. The purpose of the document's structure is to give notice of its official nature and to set up a clear space for the numbers needed for displaying the amount of tax to be withheld.

FORMAT

Format refers to the general plan or pattern of a document. It may also mean the *materials* or *topics* chosen for the document.

Chapter 5

This sample page from an employee manual displays a time clock, card, and instructions.

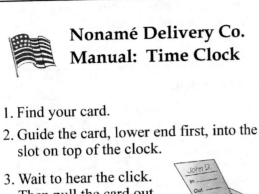

Nonamé Delivery Co.
Manual: Time Clock

1. Find your card.
2. Guide the card, lower end first, into the slot on top of the clock.
3. Wait to hear the click. Then pull the card out.
4. Replace your card in the rack.
5. NEVER clock another employee in or out.

Form 3

- The employee manual addresses newly hired workers and wants to make the material very clear.
- Graphics are an important tool in the instruction of both complicated and simple activities.
- Warnings are important messages to get across to new workers.
- The formatting focuses more on the company identity and step-by-step directions.

Below is a different form, one for indicating direct deposit of paychecks to a bank account.

Form 4

- The formatting of this document includes the company name, instructions to you and the employer, and personal, legal, and financial information.
- Employees must sign the form. The author's purpose includes having a clear, legal form to permit the transfer of money while notifying in a small print any possible legal problems.

113

Consumer, Public, and Workplace Documents

GRAPHICS

Graphics are pictures which give visual aid to texts. **Diagrams**, **charts**, and **tables** are used in a document to clarify the meaning of text. The graphics in an employee manual showing the delivery van's dashboard will be very different from the charts and tables found in the delivery route handouts.

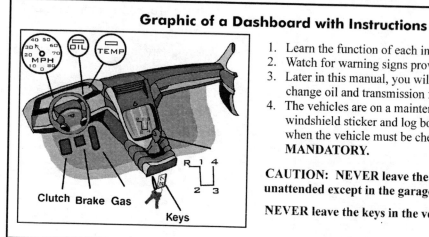

Form 5

- The graphic of the vehicle dashboard gives new employees a look at their equipment.
- The text provides information and warnings.

Delivery Route Handout

Form 6

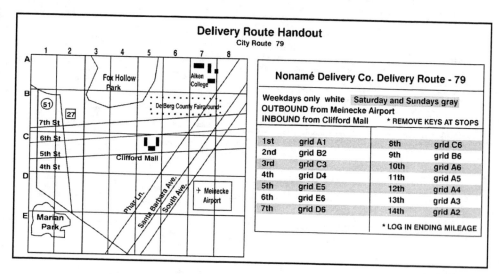

- The area map is a vital tool for drivers, who need a clear guide.
- The second part of the handout is a list of scheduled stops. The table organizes the information in an easy-to-read format.

Chapter 5

SYNTHESIZING THE INFORMATION

Imagine that an employee responsible for the deliveries on the preceding page received this map and route schedule. Then, on the bulletin board, the driver saw the notice posted at right.

The driver would need to **synthesize** the information from both the delivery route and the street renovation notice; doing so would show that the deliveries in Grid B6 on the 9th would not be possible due to the detour. The driver would then need to decide the best course of action, such as rescheduling the deliveries for the next day, when the street is open again.

HEADINGS

Headings refer to the way sections of a text are set apart by different sized or different types of fonts. The biggest heading is usually the document's title. Any following sections may have equally-sized headings or smaller ones, depending on the importance of the information. The author of the document decides how headings should be set for understanding.

Detour Notice

The Lund Department of Public Works will perform maintenance and repairs on 7th Street along the perimeter of DeBerg County Fairgrounds beginning on March 5th. A detour will reroute all traffic to 6th Street for the remainder of the week. 7th Street will re-open on March 10th.

Any inquiries should be directed to Adella Christof, DOPW Manager, at 555-7864.

Lund DOPW GTB Msg #4502860-0301004

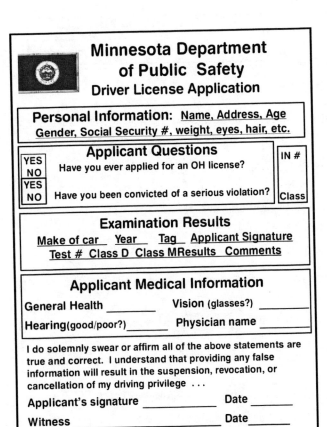

Form 7

Sample Driver's License Application (<u>not</u> an actual copy)

- The largest heading, at top, reflects the importance of which state has issued the form.

- This form is similar to other government forms that are information-gathering tools. States may vary the heading size for the different sections, but section headings are generally similar.

- The form's purpose is focused on collecting information from applicants and recording their driving test scores.

Consumer, Public, and Workplace Documents

Example of a Job Application

- Different employers look for different qualities in an employee, so forms vary widely by employer.
- In this application, the company name is most important. Next are the legal identity numbers of the applicant and the date of application.
- The remaining information has similar headings; **references** and **skills** are given plenty of space on the page. This indicates they are very important to this employer.

Practice 1: Workplace Documents

1. Look back at Form 1 on page 111. What is the *best* explanation of the author's purpose for this document, judging by the headings?

   A. The headings display a general focus on the blanks.
   B. The headings indicate the importance of the employer's name and address.
   C. For instructional purposes, the headings are presented as questions.
   D. The headings focus on the form's title and number and date, so the intended audience will have a clear idea of its importance and time frame.

   Form 8

2. Look back at Form 2 on page 112. The document requests the *most* information on _____.

   A. income tax policies for company clients
   B. employee deductions which may affect taxes in a given year
   C. how much responsibility the government has to employers
   D. new employees responsibilities to file tax forms

3. Look back at Form 3 on page 113. Which of the following statements *best* describes the structure of the document?

   A. The document has an information-gathering structure, which includes personal as well as legal identity requirements.
   B. In structure, the document presents procedures with information and an instructional diagram.
   C. The structure includes the basic graphics of an information gathering text.
   D. Documents arranged with this structure have several functions and are used to communicate with outside organizations.

Chapter 5

4. Look again at Form 3. Which of the following statements *is the best conclusion* about the format of the document?
    A. The format is simple and designed for company use only.
    B. Graphics would have helped the format of this document explain its purpose.
    C. The format is complicated and lacking in purpose.
    D. A clearer heading would help the format maintain focus.

5. Look back at Form 8 on page 116. Which of the following *best* expresses the author's purpose for this document?
    A. The document's purpose is to instruct new employees on legal issues.
    B. It will test former employees' knowledge of company policy.
    C. It will keep careful records of job applicants' personal and legal information.
    D. This document's purpose is to decide which employees will receive merit raises.

THE JOB APPLICATION PROCESS

Step 1. Prepare your responses before filling out the application. Most applications request the same basic information. Gather the information in advance and have it typed or neatly written on one page. Once you are at the employer's office, you can copy the information directly, saving you time and effort.

Step 2. List the names of your employers, your job titles, work addresses and phone numbers, and the names of your former supervisors.

Step 3. Summarize your salary and education histories. List your prior positions, with starting and ending salaries for each. Make note of additional compensation such as raises. List the names and locations of high schools and colleges you have attended. Also list any military service.

> **Example:** *Michael's Hardware Store - $9.75/hr*
> *John's Glassworks - $11.50/hr*

Step 4. Employers will ask for a listing of duties and responsibilities for each position. Rather than trying to remember and jotting down what first comes to mind, prepare two or three powerful phrases for each position.

> **Example:** *While working for Lowe's in the carpet department, I organized all the glues and cleaners into one area. With the permission of my supervisor, I organized endcap carpet accessories, so customers could find them easily.*

Step 5. Don't forget to include volunteer positions and school jobs, such as editing the newspaper, yearbook, or fund raisers. List any awards or prizes you might have won in school or at your previous job.

> **Example:** Employee of the Month, Honor Roll, Beta Club, Perfect Attendance

Step 6. List personal and professional references, giving names, addresses, and phone numbers.

Step 7. **Honesty counts.** Employers have every right to verify the information on the application before or after they hire you. Lying can result in disqualification or termination. If you had

Consumer, Public, and Workplace Documents

a dispute elsewhere, take the opportunity to give an explanation and what you learned from it.

Example: While working at the glass company, I forgot to lock down the glass after I loaded it on the delivery truck. It fell over and broke during delivery. I told my boss that it was my fault, and I would be more careful in the future. He thanked me for being honest and deducted my pay at the wholesale cost of the glass. I felt better about it, and my boss appreciated my honesty. I became more conscientious of following through with each assignment given to me.

Step 8. Fill out every field that is required. Be careful not to write in areas marked, "Office Use Only." Ask for clarification if a question does not make sense. Don't be in a hurry.

Step 9. Review the application carefully before turning it in. Check for any grammatical or spelling errors, and make sure the appropriate, businesslike tone is used throughout.

Step 10. Be sure to sign and date the application in the appropriate place.

Step 11. If you are completing an application just prior to an interview, arrive 15 – 20 minutes early, so you can take your time to fill in the form neatly. Attach a résumé to the application if you have one.

Step 12. If you picked up an application and are submitting it in order to be considered for an interview, return the completed application quickly, within 24 hours. Write a brief cover letter to turn in with your application. If at all possible, drop the application off in person. **Appearance counts.** Dress neatly, in case you get to meet the manager or department supervisor. As a guideline, wear the same thing to a job application return that you would to a church, wedding, or semi-formal dance.

Practice 2: Completing a Job Application

1. In Step 8 of "Completing a Job Application," the word *clarification* means
   A. make easier to understand.
   B. hard to understand.
   C. money.
   D. make easier to write.

2. Based on the information in Steps 3 and 6 of "Completing a Job Application," which of the following would be LEAST effective to include on the application?
   A. Employer: Financial Services; Supervisor - Jim Jones
   B. 9th grade English - grade of C
   C. Personal reference: Suekarol Baron 770-098-7654
   D. True Green Waste Department, 678 Farm Road, Wayzata, Minnesota

3. What would happen if a person gave a false name as a reference on an application?
   A. Probably nothing; they rarely call everyone you put down as a reference.
   B. He would be fined by the company for falsifying information.
   C. More than likely, he would be disqualified or terminated.
   D. They would take him out back and beat him up.

4. The prefix "-dis" helps you know that the word *disqualified* means
   A. overqualified.
   B. removed from eligibility.
   C. under qualified.
   D. being fully qualified.

5. Which word can be substituted for *appropriate* in step 10 without changing the meaning of the sentence?
   A. stolen
   B. suitable
   C. dignified
   D. wrong

Chapter 5 Summary

All **public** and **private** documents need to be filled out and proofread carefully, and knowing what to expect from each will save time, money, and stress down the road. Forewarned is forearmed when it comes to government and business documents, so have all personal information close at hand when you fill them out.

To **synthesize** information is to take data from two separate sources and combine them into a new and original whole. Synthesizing is essential when making decisions based on procedure and real-world events. Synthesizing helps to draw conclusions.

Consumer documents are forms and materials used in the purchase, maintenance, and use of products or services. They include warranties, information guides, and instruction manuals.

Personal documents include both manuals and information-gathering forms. They include personal and professional licenses, Social Security cards, and insurance forms.

Public documents are forms and documents available to help citizens in ordinary business and errands. They include bank statements, loan documents, and travel and communication information.

Structure means the way parts of something are arranged. Most documents have sections for personal information and for explanations of the document itself.

The word **format** means the general plan or pattern of a document. It may also mean the *materials* or *topics* chosen for the document.

Graphics are pictures which give a visual aid to texts. **Diagrams**, **charts**, and **tables** are used in a document to clarify the meaning of text.

Headings refer to the way sections of a text are set apart by different sized fonts or different types of fonts. The biggest heading is usually the document's title.

The **job application process** involves giving detailed, accurate information about yourself, your education, and your work experience in a way that is neat, concise, and easy to understand.

CHAPTER 5 REVIEW

Read the following informational material, and answer the questions that follow.

1. What is the BEST description of the purpose of this document?

   A. It is a confusing document that instructs people on how to apply for a Social Security number.

   B. The author's purpose is that the material can be used as a workplace document helping new employees understand Social Security numbers.

   C. Its purpose is to be an information gathering government document that is an application for a Social Security number.

   D. The author's purpose is to use the material as a public document or product.

2. What can you conclude about the structure of this document?

   A. The structure is simple and made for public use so citizens can apply for a Social Security card without going through the government agency.

   B. A smaller heading would help the structure flow evenly and give equal importance to other parts of the document.

   C. Charts would have helped the structure of this document focus on its purpose and audience.

   D. The structure is complicated, but it is well organized by numbering and grouping the material.

3. Which of the following statements BEST describes the document headings?

   A. The headings are typical of a government document: the title is largest, and all other headings are equal to each other, except for optional information.

   B. The headings are all the same; therefore, all sections are similar in importance and applicants must completely fill them out.

   C. The document's headings help display the meaning of the graphics in an organized and structured form.

   D. The sections have all different type headings since the document's purpose is to make the report easy for government office employees to fill out.

Consumer, Public, and Workplace Documents

4. The format is BEST described by which of the following statements?
   A. The document is informative in format: it has all the information that a person will need to know to apply for Social Security benefits.
   B. The author's purpose in the formatting of the document was to leave space for projected benefits.
   C. The format follows the usual company material, needing to know the employee's name and work area.
   D. The document is information-gathering in format: it asks for a fairly detailed history of the person applying for the Social Security card.

Tracy has been hired by Yap, Inc. On her first day, she will need to fill out both a W-4 form and a direct deposit application.

5. What will Tracy need to bring with her to fill out these forms completely?
   A. her driver's license and bank account number
   B. her student identification card and Social Security card
   C. her Social Security card and a void check or bank letter
   D. her bank account number and a checking deposit slip

6. _____ refers to the charts, tables, and diagrams found in a text.
   A. Headings   B. Artwork   C. Graphics   D. Public documents

7. The tax form filled out by employees prior to starting work is the _____ form.
   A. W-4
   B. social security identification number
   C. loan documentation
   D. binding contract

8. _____ documents include bank statements, travel information, and loan documents.
   A. Consumer   B. Public   C. Personal   D. Government

9. _____ documents include warranties, instructions, and government information guides.
   A. Consumer   B. Public   C. Personal   D. Government

10. _____ documents include driver's licenses, Social Security cards, and passports.
    A. Consumer   B. Public   C. Personal   D. Government

Chapter 5

WEB SITES

www.consumer.gov

A brightly decorated and sectioned site that has consumer information, forms, and links to more forms of all kinds: school loans, housing, child care loans, etc. There are both federal and individual state government forms. Some of the features include a Consumer Action Handbook, Fraud information, and current consumer news alerts.

http://www.ifg-inc.com

Interesting look to this site. Careful though. It does have several ways for you to get into a "buying on line" situation. The features include links to federal, state, and private consumer agencies and Web sites. There is consumer information on cars, travel, education, etc.

www.lawsmart.com

This site definitely has a tremendous amount of legal forms and information. However, it can be difficult to pin down what you want to see and/or print. I worked steadily for almost an hour to download a promissory note from the HUD page. There is legal information on this site for both federal and state law.

www.consumerworld.org

Another great, nonprofit consumer information site. Includes information from Consumer Reports magazine and an easy-to-use site search engine.

http://www.thesite.org/workandstudy/gettingajob/applications/applicationforms

The somewhat complicated URL is redeemed by the amount of information and practice contained on the site. It's non-profit, easy to use, and includes detailed instruction and common-sense advice.

Consumer, Public, and Workplace Documents

MCA-II/GRAD Reading Test
Practice Test One

The purpose of this test is to measure your progress in reading comprehension skills. This test is based on the Minnesota standards for English and Language Arts and adheres to the sample question format provided by the Minnesota Department of Education.

General Directions:

1. Read all directions carefully.

2. Read each question or sample. Then choose the best answer.

3. Choose only one answer for each question. If you change an answer, be sure to erase your original answer completely.

4. After taking the test, you or your instructor should score it using the answer key that accompanies this book.

Practice Test One

> Though he played in the Negro Leagues in the 1940s and 50s, Buck O'Neill never made it to the Baseball Hall of Fame. Read about the life of this outstanding player. Then answer question 1 – 8.

Buck O'Neil

1. John Jordan "Buck" O'Neil was a first baseman and manager in baseball's Negro leagues during the thirties, forties and fifties. He is best known for his playing career with the Kansas City Monarchs of the Negro leagues and as a coach and scout for the Chicago Cubs and Kansas City Royals in the major leagues.

2. O'Neil was born Nov. 13, 1911, in Carrabelle, Florida. Due to racial segregation, he was denied the opportunity to attend high school in Sarasota, Florida, or play baseball in the major leagues. He began his baseball career with the Memphis Red Sox of the Negro leagues in 1937 and was traded to the Monarchs the following year. A tour of duty in the Navy during World War II briefly interrupted his playing career.

3. In the Negro leagues, O'Neil amassed a career batting average of .288, including four .300-plus seasons at the plate. In 1946, the first baseman led the Negro League in hitting with a .353 average and followed that in 1947 with a career best .358 mark. He posted averages of .345 and .330 in 1940 and 1949, respectively.

4. In 1948, he took over as manager of the Monarchs and guided the team to league titles in 1951, 1953, and 1955. He played in four East-West All-Star games and two Negro League World Series during his playing days.

5. O'Neil also joined the legendary Satchel Paige as a teammate during the height of the Negro League barnstorming days of the 1930s and 1940s. This was a period when teams of Negro all-stars would travel the countryside playing town teams, college teams, and teams of major leaguers to earn extra money and gain exposure for the Negro leagues.

6. O'Neil left the Monarchs following the 1955 season, and in 1956, became a scout for the Chicago Cubs. In 1962, the Cubs hired him as a coach, making him the first black coach in Major League Baseball history. As a scout, he signed Hall of Fame outfielder Lou Brock to his first pro contract.

7. He is sometimes credited with having signed future Cubs' Hall of Fame second baseman Ernie Banks to his first pro contract, but in fact, only signed him to his first MLB contract. Banks had actually been scouted and signed to the Monarchs by Cool Papa Bell, manager of the Monarchs' barnstorming "B" team, in 1949. Banks played for the Monarchs in 1950 and briefly in 1953 when O'Neil was his manager.

8. O'Neil has worked as a Kansas City Royals scout since 1988 and was named Major League Baseball's "Midwest Scout of the Year" in 1998.

9. O'Neil gained national prominence during the late 1990s with his poignant and compelling narration of Negro league history as part of Ken Burns' PBS documentary on baseball. He has since been the subject of countless national interviews, including appearances on *Late Night with David Letterman* and the *Late, Late Show with Tom Snyder*.

10. Today, Buck O'Neil serves as honorary Board Chairman of the Negro Leagues Baseball Museum (NLBM) in Kansas City, Missouri. He was a member of the 18-member Baseball Hall of Fame Veterans Committee from 1981 to 2000 and was instrumental in the induction of eight Negro League players during that time.

11. O'Neil was a candidate in 2006 for induction into Major League Baseball's Hall of Fame in a special vote for Negro League players, managers, and executives. However, he did not receive the necessary nine votes for induction from the 12-member committee.

12. "God's been good to me," O'Neil commented to 200 well wishers after hearing that he had not been elected to the Hall at age 94.

13. "They didn't think Buck was good enough to be in the Hall of Fame. That's the way they thought about it, and that's the way it is, so we're going to live with that."

14. "Now, if I'm a Hall of Famer for you, that's all right with me. Just keep loving old Buck. Don't weep for Buck. No, man, be happy, be thankful."

1. The theme of the passage can best be summarized by which of the following?   I.D.6, I.D.10, I.C.G11
   A. O'Neil personifies what is wrong with baseball today.
   B. O'Neil probably deserves to be in the Hall of Fame.
   C. O'Neil is a great ambassador of good will and a good family man.
   D. Baseball is better off with people like Buck O'Neil representing it.

2. The last two paragraphs tell us what about O'Neil's outlook on life and his character?   I.C.7, I.C.G12
   A. He is bitter about not being let into the Hall of Fame.
   B. He feels his life is not yet complete.
   C. He is a positive man and doesn't let things get him down.
   D. He is a happy man, but would be happier if allowed into the Hall of Fame.

3. The author's purpose in writing the passage is most likely?   I.C.6, I.C.9
   A. to introduce an interesting person
   B. to inform the reader of an injustice in the world
   C. to persuade the reader to be active in a cause
   D. to relate an adventure

4. Which paragraph in the passage is unnecessary in detailing the key points of O'Neil's life?   I.C.6, I.C.9
   A. paragraph 11, about his current life
   B. paragraph 4, about his baseball statistics
   C. the last paragraph
   D. paragraph 8, about the signing of Ernie Banks

5. What can the reader infer from O'Neil's experiences about baseball in the first half of the 20th century? I.C.7, I.C.G12
   A. Negro league players were better than major league players.
   B. Baseball was not very popular during that time.
   C. Baseball did not allow Negroes to play in the major leagues.
   D. Most of the baseball teams were in the Midwest.

6. The word "poignant" as used in describing O'Neil's narration of the Negro leagues most closely means I.B.2, I.B.G6
   A. interesting and exciting.
   B. troublesome.
   C. touching and memorable.
   D. shocking, hard to believe.

7. The story of O'Neil is told from the point of view of I.C.6, I.D.4
   A. a narrator.
   B. O'Neil himself.
   C. first person.
   D. second person.

8. The author of this selection is likely a I.C.G12, I.C.8
   A. former player.
   B. baseball historian.
   C. friend of O'Neil's.
   D. fiction writer.

The Cherokee Indians protested when the United States government broke their treaties. Read their cry for justice. Then answer questions 9 – 18.

Cherokee Indians Removal to Oklahoma

We wish to remain on the land of our fathers. We have a perfect and original right to claim this, without interruption or molestation. The treaties with us…guarantee our residence, and our privileges, and secure us against intruders. Our only request is that these treaties may be fulfilled, and these laws executed.

But if we are compelled to leave our country, we see nothing but ruin before us. The country west of the Arkansas territory is unknown to us. From what we can learn of it, we have no good news in its favor. All the inviting parts of it, as we believe are preoccupied by various Indian nations, to which it has been assigned. They would regard us as intruders, and look upon us with the evil eye. The far greater part of the region is, beyond all controversy, badly supplied with wood and water; and no Indian tribe can live as agriculturalists without these articles. All our neighbors, in case of our removal, though crowded into our near vicinity, would speak a language totally different from ours, and practice different customs. The original possessors of that region are now wandering savages, lurking for prey in the neighborhood. They have always been at war, and would be easily tempted to turn their arms against peaceful emigrants… It contains neither the scenes of our childhood, nor the graves of our fathers…

Shall we be compelled by a civilized and Christian people, with whom we have lived in a perfect peace for the last forty years, and for whom we have willingly bled in war, to bid a final adieu to our homes, our farms, our streams and our beautiful forests? No. We are still firm. We intend still to cling, with our chosen affection, to the land which gave us birth, and which, every day of our lives, brings to us new and stronger ties of attachment. We appeal to the judge of all the earth, who will finally award us justice, and to the good sense of the American people, whether we are intruders upon the land of others. Our consciences bear us witness that we are the invaders of no man's rights — we have robbed no man of his territory — we have usurped no man's authority, nor have we deprived anyone of his unalienable privileges. How then shall we indirectly confess the right of another people to our land by leaving it forever? On the soil which contains the ashes of our beloved men we wish to live — on this soil we wish to die.

9. What concerns did the Cherokee raise about the land itself they were being forced to move to?
   A. There was no grass for their animals to graze on.
   B. There were hostile bands of Indians rampaging the countryside.
   C. It lacked sufficient resources for them to farm and grow food.
   D. The water was polluted.

10. To whom were the Cherokee ultimately pleading?
    A. to congress
    B. to God and to the American people
    C. to the Indian nations
    D. to the Army who were forcing them to move

11. What was the basic argument the Cherokee made for being allowed to stay where they were and <u>not</u> move?
    A. They had invaded nobody's rights, so why should their rights be invaded.
    B. They owned the land already, so should not have to move.
    C. The land was not fit for anyone but them to live on.
    D. Their ancestors had guaranteed them the land forever.

12. What does the phrase "indirectly confess the right to another people to our land" most probably mean?
    A. By moving, they would be admitting that they could not maintain the land.
    B. By confessing it was not their land, they would be shamed.
    C. By admitting it was not their land, they would be seen as thieves.
    D. Giving their land to another people would be admitting it was not truly their land.

13. What did the author cite as the fundamental reason they should not have to leave their land?
    A. that the women and children be spared moving away
    B. the original treaty with the Cherokee
    C. it will dishonor their ancestors to move
    D. living on someone else's land is not the Cherokee way

14. What type of argument does the writer present in making his case for staying?
    A. He uses statistics and examples to make his point.
    B. He appeals to the sentiment or good nature of the reader.
    C. He used legal language to prove his argument.
    D. He dismisses the argument as unimportant.

Practice Test One

15. Who is most likely the author's audience?   I.C.7, I.C.G12
    A. the president of the United States
    B. the Indian nation
    C. the military leader sent to force the Cherokee to move
    D. the war chief of the neighboring tribe

16. What evidence does the author give that in the new place, his people will not be welcome?   I.C.6, I.D.14
    A. It is already occupied by other Indian nations, some at war already, and with most having different customs and languages than the Cherokee.
    B. He says that the people living there have ravaged the land and made it unfit to live on by any newcomers.
    C. He details how the land is unfit for farming and incapable of maintaining a good supply of food.
    D. He exaggerates the difficulty of moving his entire nation of people such a long distance.

17. What does the word "usurped" near the end of paragraph three most closely mean in this passage?   I.B.2, I.B.G6
    A. sugary, slimy
    B. stolen or taken away
    C. slippery to the touch
    D. humiliated, put down

18. In two or three sentences, show evidence of how determined the Cherokee are to stay on their land and not move away. Use your own sheet of paper.   I.D.14

As the time approaches for someone to start college, there are many decisions to be made. Read the following discussion about choices for housing. Then answer questions 19 – 28.

College Decisions: Living at Home vs. Living Away

When a young person begins to make a decision about which college to attend, there are many factors to consider. Students today find that one of the first decisions they are faced with is the question of whether they want to stay at home and commute to a local college — if one is nearby and offers the courses a students is interested in — or live away from home in a dormitory or apartment. There are several pros and cons for each choice. Students should think carefully about their personality, finances, and family situation when deciding which option is more suitable to them.

The cost of housing is very often the primary consideration. If you live at home and commute, you will not pay rent and food costs like you would if you lived in an apartment or dormitory. If you will be paying your own college fees or using student loans, this can be an appreciable amount of money saved. Although many students help out their family by paying room and board, probably it will not be as expensive as living away from home.

Likewise, transportation is an important consideration. If you choose to commute, you need either public transportation, a vehicle of your own, or a dependable arrangement for getting to and from campus. However, if you have to have your own vehicle, you may want to calculate to see if driving is cheaper or a better use of money than living on campus and using the school transportation system. Purchasing and maintaining a vehicle may be as expensive as living on campus in the long run.

While living on your own gives you a great amount of freedom and helps you learn to be independent, living at home has the advantage of being familiar and providing structure, allowing you to better focus on your studies. Many students living on their own for the first time find they are overwhelmed by their new freedom and the responsibilities that go along with it. Still, while you may give up some independence choosing to live at home, commuting does require extra preparation and proper organization. You can't run back to your house across town for books or other forgotten items; you have to plan ahead for what you will need on campus each day.

Your individual family situation also could play an important part in the decision. Are you needed at home to help out with family responsibilities, a family business, elderly grandparents, or younger siblings? Do you need to stay at home because your family depends on you? I had to weigh these things against living on campus when I was deciding on how to handle college, and it was not an easy choice for me. However, if other arrangements can be made, being away from those commitments should be beneficial in allowing you to concentrate more on your studies.

19. Which side of the argument does the author most likely fall on?
    A. living at home, but taking public transportation
    B. living in a dormitory on campus
    C. living at home and commuting
    D. choosing a college in another city

20. In paragraph two, the word "appreciable" most likely means
    A. very little.
    B. outrageous.
    C. something appreciated.
    D. considerable or substantial.

21. The main idea of the passage is:
    A. There are many considerations when deciding on a college.
    B. College can be very expensive and expenses should be calculated down to the penny.
    C. Each person has his own reasons for staying at home and commuting to college.
    D. The newfound freedom of heading off to college can be very exciting.

22. What happens in the final paragraph that is inconsistent with the rest of the passage?
    A. The author switches into first person, changing the point of view of the writing.
    B. The author becomes cynical toward going to college.
    C. The author tries to strongly persuade the reader toward his viewpoint.
    D. The author moves into third person.

23. A one-sentence paraphrasing of paragraph three might read: I.C.5, I.C.G11
    A. When deciding on how to commute, a person must factor in whether they want to use their own car or someone else's.
    B. When choosing how to get to and from college, a person needs to consider whether it is more cost-effective to stay home and commute or to live on campus and get around without a car.
    C. To get around the campus, many people choose the school's transportation instead of their own.
    D. In order to make a sound decision on commuting, one has to factor in the time it takes to get from home to campus versus the time it might take to go from the dorm to class.

24. How might living on your own for the first time be "overwhelming," as the author puts it? I.C.7, I.C.G12
    A. There is so much going on, you may not be mature enough to juggle all your classes.
    B. Your studies may be too difficult for you at first.
    C. The freedom may cause some students to not properly budget their study time.
    D. Freedom from responsibilities at home may cause you to get homesick.

25. Which of the following is most likely the best qualified to write this passage? I.G.12, I.D.4, I.C.8
    A. a professor of liberal arts
    B. a high school senior looking at colleges
    C. the parent of a prospective collegian
    D. an upperclassman at a local college

26. How might being away from your family actually help you in studying? I.C.7, I.C.G12
    A. You would have a roommate with whom to attend class.
    B. You would be closer to campus.
    C. It could allow you more quiet time and fewer distractions.
    D. Your study habits would have improved since moving.

27. What type of argument style does the author use in this passage? I.C.6, I.D.4
    A. inflammatory words and tone
    B. logical, step-by-step reasoning
    C. an impassioned plea
    D. lofty claims and ideas

28. Write a one-sentence conclusion that summarizes the passage. Use a separate sheet of paper. I.C.5, I.C.G11

End of Section 1. Check your work.

Section 2

> The following two passages explore how Western ideas have become part of Asian countries. Read the passages. Then answer questions 29 – 35.

Passage 1

It is amazing how the Japanese have retained their cultural heritage while simultaneously integrating many parts of Western culture. One of the most popular adaptations is the style of dress. Many Japanese today wear "western" style clothing such as business suits, active wear, jeans and T-shirts. Traditional clothing is now often only reserved for special occasions.

Many Japanese also have adopted western furnishings into their homes. It is not unusual to have a completely westernized home with only one traditional Japanese room. Western influences can be seen throughout Japanese culture, such as fast-food restaurants, music, and the movies.

The Japanese also have more time to devote to leisure. Surveys show that spending time with family, friends, home improvement, shopping, and gardening form the mainstream of leisure, together with sports and travel. The number of Japanese making overseas trips has increased notably in recent years. Domestic travel, picnics, hiking and cultural events rank high among favorite activities.

Japan is a land with a vibrant and fascinating history, with varied culture, traditions and customs that are hundreds of years old. Still, segments of its society and economy are as new as the microchips in a personal computer.

Passage 2

India is the seventh largest country in the world with a population of nearly one billion people. Only China contains more people. It is a land of deserts, plains, jungles, and mountains. The people speak about 180 different languages and come from many different races and religious backgrounds.

Many Indian customs have remained the same for hundreds of years, even though many social and scientific advances have also occurred. For example, cows, which are sacred to many practicing Hindus, are allowed to roam freely in modern business districts. Factory workers may wear traditional costumes on the job. In rural areas, girls take care of younger brothers and sisters at home while their brothers go off to school.

For hundreds of years, India was a land of mystery, wealth, and adventure. Columbus thought he was in India when he discovered America. Other European explorers later found India and its famous jewels, rugs, silks and spices. When the British entered India in the late 1700s, they governed many parts of the country. They built roads, telephone systems, and railroads and established a system of education still in place today. Under Gandhi, India achieved independence from Great Britain in 1947. One year later, he was assassinated. Today, India exists as an independent, democratic republic.

Practice Test One

29. Passage 1 and Passage 2 differ slightly in their presentation in that: I.D.7, I.D.6, 10
    A. Passage 1 deals more with the country's people than with the country's history.
    B. Passage 2 deals more with the education of the people than Passage 1.
    C. Passage 2 deals more with the country's people than the country's history.
    D. Passage 1 is more negative about the future of the country than Passage 2.

30. Which would have been a better choice of wording the following sentence: "Western influences can be seen throughout Japanese culture, such as fast-food restaurants, music, and the movies.   I.D.4, I.C.6
    A. Western influences, fast-food restaurants, music, and movies, can be seen throughout Japanese culture.
    B. Western influences — such as fast-food restaurants, music, and movies — can be seen throughout Japanese culture.
    C. Western influence such as fast-food restaurants and music and movies can be seen throughout Japanese culture.
    D. The sentence is fine as it is.

31. Which is an example of irony found in Passage 1?   I.D.5, I.D.4
    A. The Japanese also have more time to devote to leisure.
    B. Segments of Japanese society are as new as the microchips in a personal computer.
    C. One of the most popular adaptations is their style of dress.
    D. Traditional clothing is now often only reserved for special occasions.

32. In paragraph two of Passage 2, the term "Hindus" probably most closely refers to whom?   I.B.2, I.B.G6
    A. factory workers in India
    B. people who speak several different languages
    C. religious people in India
    D. European explorers who discovered India

33. How does the author in Passage 1 do a better job of making his argument than in Passage 2?   I.C.6, I.C.G12
    A. Passage 1 uses more superlatives and descriptive adjectives than Passage 2 does to make its points.
    B. Passage 1 sticks to the subject, using relevant examples and descriptions, while Passage 2 includes several bits of non-essential information.
    C. Passage 2 doesn't tell enough about the history of the country.
    D. Passage 1 cites precise statistics in developing its argument; Passage 2 deals more in general terms.

34. Both passages are most likely written from the point of view of what type of person?   I.C.G12
    A. a person who has visited both countries
    B. a travel agent
    C. a Hindu
    D. a resident of that country

35. What similar technique is found in both passages?   I.D.6, 10, I.D.7
    A. Both passages speak about that country's customs and traditions.
    B. Both passages make a case for its being a wonderful place to live.
    C. Both passages arrive at similar conclusions about the people there.
    D. The passages are not similar at all.

The diversity in world religions is depicted in the following graphic and further described in the accompanying article. Read over them both. Then answer questions 36 – 42.

Religions of the World

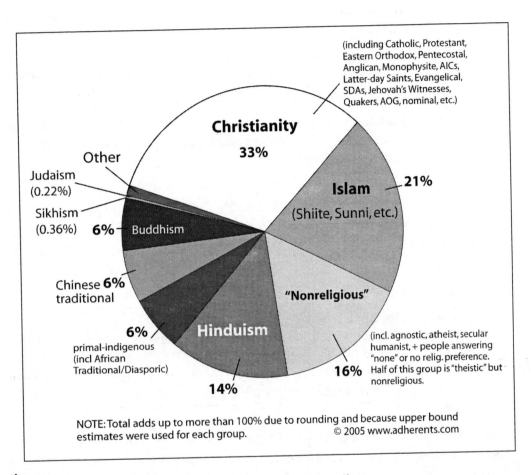

NOTE: Total adds up to more than 100% due to rounding and because upper bound estimates were used for each group.   © 2005 www.adherents.com

Introduction

The percentages above represent current estimates of the number of people worldwide who have at least some degree of identification with a specific religion. Levels of participation, of course, vary within all groups, but these numbers are reasonable estimates as to the number of "followers" of the religions listed. Different criteria and measuring methods can be used to compile such a list, but experts tend to agree on the order of the popularity of the religions shown above, if not the actual percentages.

In modern Western thought, the first writers to divide the world into "world religions" were the Christians. Originally, three religions were identified and recognized. Those religions were Christianity, Judaism and paganism (i.e., everybody else). After many centuries and with the increased Western awareness of Eastern histories and philosophies — and with the development of Islam — other religions were added to the list.

Christianity

Christianity is a monotheistic religion centered on the life, teachings, and actions of Jesus of Nazareth — known by Christians as Jesus Christ — as told in the writings of the New Testament in the Bible.

Christianity's origins are intertwined with Judaism, with which it shares much sacred text and early history. Specifically, it shares the Hebrew Bible, known in the Christian context as the Old Testament.

Estimates of how many followers there are of Christianity are about 1.9 billion (or about 33% of the world population). Regardless of the degree of accuracy of this figure, Christianity, if taken as a whole, is inarguably the largest world religion.

It should be understood, however, that Christianity is an umbrella term that comprises many different branches and denominations, such as Catholicism and Protestantism and their many sub-denominations.

Islam

Islam is a monotheistic faith and the world's second-largest religion. Followers of Islam are known as Muslims. Muslims believe that God revealed his divine word directly to mankind through many prophets and that Muhammad was the final prophet of Islam.

Contemporary figures for Islam are usually between 900 million and 1.4 billion, with 1 billion being a figure frequently given in comparative religion texts.

Secularism (and the non-religious)

Secularism is commonly defined as the idea that religion should not interfere with or be integrated into the public affairs of a society. It is often associated with the Age of Enlightenment in Europe and plays a major role in Western society. The principles, but not necessarily practices, of separation of church and state in the United States draw heavily on secularism.

One portion of this broad grouping includes those who are best described as "non-religious." This group is made up of those who are essentially passive with regards to organized religion, generally affirming neither belief nor disbelief. They may be neither contemplative about philosophy and spirituality nor involved in a religious/faith/philosophical community.

Although a certain percentage of people in many countries classify themselves as non-religious in surveys, many of these have their own personal philosophy but have no stated affiliation with any organized religion. Conservative estimates calculate a little less than one billion people worldwide can be classified as "non-religious" (about 15.2% of the world).

Hinduism

Hinduism is described as a diverse body of religion, philosophy, and cultural practice native to and predominant in India. It is characterized by a belief in reincarnation and a supreme being of many forms and natures, by a view that opposing theories are aspects of one eternal truth, and by a desire for liberation from earthly evils.

The highest estimates compiled for the number of people worldwide who practice

Hinduism is about 1.4 billion, which would actually place it higher in participation than Islam. A closer figure for Hinduism is generally accepted at around 850 million to 900 million people.

Judaism

Judaism is the religion of the Jewish people, with around 14 million followers as of 2005. It is one of the first recorded monotheistic faiths and one of the oldest religious traditions still practiced today.

Judaism in all its variations has remained tightly bound to a number of religious principles, the most important of which is the belief in a single, omniscient, omnipotent God who created the universe and continues to be involved in its governance.

According to Jewish thought, the God who created the world established a covenant with the Jewish people and revealed his laws and commandments to them in the form of the Torah.

The practice of Judaism is devoted to the study and observance of these laws and commandments.

36. Which two of the religions described in the passage are said to be closely related?
    A. Christianity and Judaism
    B. Judaism and Hinduism
    C. Christianity and Islam
    D. Islam and Buddhism

37. Christianity, Judaism, and Islam are all described as being "monotheistic." What is most likely the meaning of monotheistic?
    A. being the dominant religion in a country
    B. belief in eternal truths and one set of rules to live by
    C. belief in one God or one Supreme Being
    D. being separate from other influences

38. According to the chart, which of the five defined religions has the fewest followers today worldwide?

    A. Christianity   C. Hinduism
    B. Judaism   D. Secularism

39. People who do not believe in any God or higher being — atheists — would probably fall into which of the religious categories listed?
    A. Judaism   C. Islam
    B. Hinduism   D. Secularism

40. The top two religions account for about how many of the people in the world who ascribe to some religion?
    A. over half
    B. less than one-fourth
    C. less than half
    D. almost three-fourths

Practice Test One

41. The author's purpose for this article is to

    A. educate.
    B. cause doubt.
    C. entertain.
    D. persuade.

42. What is the probable reason for Judaism being included in the passage when it has such a small following?

    A. It is popular in Europe.
    B. The author may ascribe to Judaism.
    C. It was once among the top three religions in the world.
    D. Christianity and Islam share its monotheistic beliefs.

> The French were on the U.S. side in the Revolutionary War. Read the following article about how they helped us defeat the British. Then answer questions 43 – 52.

A French Revolution

The Battle of Yorktown, which took place in October of 1781, was the turning point in the American Revolutionary War. It was also the battle that involved the greatest amount of help from the French. In fact, of the four major leaders on the side of the Americans during the Battle of Yorktown, three were French. General George Washington was the fourth. The French certainly influenced the outcome of the Battle of Yorktown, and the army that fought against the British was a Franco-American army. However, some of the French soldiers were not happy to help the Americans.

The battle would not have taken place as it did had it not been for a French military leader, Admiral Rochambeau. When he met with Washington in 1781, Washington was about to attack New York to win it back from the British. However, Rochambeau convinced Washington that it would be wiser to go south to Yorktown. He had heard from another French general, Lafayette, that the British general Cornwallis had taken Yorktown. Because of this news, Rochambeau advised Washington to deal with Cornwallis and forget the British in New York for the time being. Washington took Rochambeau's advice, and the two of them, along with their troops, marched south.

Meanwhile, Rochambeau notified yet another French officer, Admiral Françoise de Grasse, to sail into Chesapeake Bay, near Yorktown, and to confront the British naval vessels that were there. De Grasse did so and defeated the British fleet. As a result, Cornwallis had no naval back-up to help him hold Yorktown.

After that, de Grasse's men, Rochambeau's men, and Lafayette's men all joined with Washington's forces in laying siege to Yorktown, surrounding Cornwallis' men and starving them for 21 days.

On September 19, 1781, Cornwallis surrendered to the French and American forces. As a result of this defeat, the British government began to realize that winning the war against the American colonists would be next to impossible. Because of this realization, the prime minister of Great Britain immediately resigned. This resignation led to the end of the American Revolutionary war.

The help of the French, who had long admired the revolutionary spirit of the America colonists, made a crucial difference in the course of the war for American freedom.

43. The tone of the first paragraph lets the reader know what about the author's feelings?   I.D.5, I.C.G11
    A. He dislikes the French people.
    B. He is pro-American.
    C. He doesn't think much of the French leaders.
    D. He is pro-French.

44. Which of the following is an opinion, not a fact?   I.C.7, I.C.G12
    A. On September 19, 1781, Cornwallis surrendered to the French and American forces.
    B. Washington marched south with his troops toward Cornwallis.
    C. The Battle of Yorktown was the only turning point in the American Revolution.
    D. Cornwallis had taken Yorktown.

45. What is the main idea of the passage?   I.C.5, I.C.G11
    A. It is best to get advice from other military leaders when fighting a war.
    B. Many of the American officers in the Revolutionary War were inferior.
    C. Much of the fighting in the Revolutionary War took place near New York.
    D. The French did a great deal to help the United States in the Revolution.

46. Why was going to Yorktown a better idea than attacking New York?   I.C.7
    A. Cornwallis was there and needed to be defeated first.
    B. The British were easier to defeat than Cornwallis.
    C. Getting to Yorktown would be easier than getting to New York.
    D. Cornwallis was in New York and would be difficult to defeat.

47. The term Franco-American most likely refers to   I.B.2, I.B.G6
    A. the French being first in helping the Americans.
    B. a mixture of French and Americans.
    C. the French being better than the Americans.
    D. delicious lunchtime pasta.

48. What reason is given for the French helping the Americans?   I.C.7
    A. The French admired the Americans' spirit.
    B. The French realized the Americans had poor information.
    C. The French leaders felt superior to the Americans.
    D. The Americans pleaded for help from the French.

49. What kinds of words does the author use to let the reader know this is more likely a narrative than a straight non-fiction piece?   I.D.4
    A. He changes some of the facts of the actual events.
    B. His comments about Americans are more negative than his comments about the French.
    C. He uses several superlatives and descriptive adjectives in the passage.
    D. He injects humor in the passage to make his point.

50. Considering strictly the events outlined in the passage, what was the most valuable assistance the French provided the colonists?   I.C.7 I.C.G12
    A. leadership and important information
    B. their expertise in fighting naval battles
    C. their large, experienced army
    D. clothing and much-needed weapons

Practice Test One

51. Which sentence below from the passage actually weakened the author's argument for the French's involvement in helping the Americans? I.C.6 & 9 I.C.8

    A. "As a result, Cornwallis had no naval back-up to help him hold Yorktown."
    B. "However, some of the French soldiers were not happy to help the Americans."
    C. "Because of this realization, the prime minister of Great Britain immediately resigned."
    D. "However, Rochambeau convinced Washington that it would be wiser to go south to Yorktown."

52. Describe in a short paragraph what changes in style the author could have made to make the passage more objective. Use a separate sheet of paper. I.D.14 & 9 I.D.7

Following directions for new camera equipment can be a challenge. Read the following set of directions. Then answer questions 53 – 57.

To Make Photos for Publication

1. Connect the camera to any computer using a "Y" coaxial cable.
2. Photo program will automatically load.
3. Find photo(s) to be used in "Recent Shot" folder.
4. Highlight the desired photo(s) by clicking once.
5. Drag the photo(s) to your existing picture folder.
6. Open Photoshop.
7. In Photoshop, open your picture folder and select one photo.
8. Crop and size the photo to fit the desired space on the page.
9. Adjust the color format by using the "Auto Select" feature.*
10. Adjust the gradation levels by using the "Auto Grade" feature.
11. Adjust sharpness (up or down) with the "Filter" feature.
12. Lighten the photo by holding down the CTRL+M keys. Lighten to 75%.
13. Repeat for subsequent photographs, as needed.

* Note: For black and white photos, choose "Eliminate Color" in Step 9.

Always make sure your equipment is clean and free of smudges and smears. Be sure to wash your hands before using any company equipment. All persons taking a camera off the premises should sign for the camera with the inventory clerk in Room 120. Always return any and all equipment used to the inventory clerk as soon as you return from an assignment. Please be sure to replace all film or batteries used.

Our policy is, and always will be, to take care of our equipment so as to take the finest photographs in the business. Take pride in your work, and your work will make you proud.

53. These instructions are most likely instructions for how to I.C.3
    A. make photographs look better on your computer.
    B. prepare a computer slide show for later use.
    C. prepare photographs for printing in a newspaper or magazine.
    D. move photographs into a folder for safekeeping.

54. The term "Publication" in the heading most likely refers to I.C.3, I.B.2
    A. how to set up a camera for good photographs.
    B. how to properly prepare the photos for maximum reproduction quality.
    C. ways to get in position to photograph well and in good light.
    D. how to ready the camera for anything that might be photo worthy.

55. Which of the following statements best describes the document's style? I.C.3, I.C.7
    A. formal and designed to simplify a process
    B. informal and designed to casually instruct
    C. semi-formal and non-threatening
    D. formal, but with a distinct tone of humor

56. The graphic was included most probably as an attempt to I.C.3, I.D.14
    A. show what a finished picture would look like.
    B. break up the wordiness of the piece.
    C. distract the reader from the serious nature of the message.
    D. add a visual aid to help soften the formality of the piece.

57. "Coaxial" cable is most likely a I.C.3, I.B.2
    A. special cable used to connect auxiliary equipment to a computer.
    B. special cable used to reproduce photographs.
    C. special cable used by a photo editor to monitor picture quality.
    D. special cable used to load Photoshop onto a computer.

End of Section 2. Check your work.

> Do animals have feelings? See what the following article has to say about it. Then answer questions 58 – 65.

Animal Emotions

Do elephants cry? Can whales fall in love? Can chimpanzees die of a broken heart? Recent evidence suggests that animals may actually experience a variety of emotions ranging from fear and aggression to love, sadness and joy. In fact, animals and humans share not only common feelings, but also a similar brain anatomy and chemistry.

For many years, scientists believed that animals displayed only instincts, such as the impulse to flee from predators or the urge to attack intruders. Scientists called these instincts primary emotions because they required no conscious thought. Scientists denied that animals could experience higher emotions such as happiness, sadness or jealousy. One notable exception to these beliefs was the famous scientist, Charles Darwin. Darwin insisted that humans and animals share common emotional links. However, few scientists took his assertions seriously.

Unlike earlier experiments with caged animals, today's scientists frequently watch animals in their natural settings. They conduct field studies that involve careful observations over long periods of time. Researchers take detailed notes on what they see and hear, often photographing or videotaping the variety of animal behaviors they see. They then draw conclusions based on their observations, often sharing what they have learned through television programs on the *Discovery Channel* and *Animal Planet* and in magazines like *National Geographic.*

As a result of observing animal behavior in the wild, researchers have now discovered that animals display an array of emotions. Yes, elephants do cry. Or at least they appear to, especially when an old elephant dies. The other elephants stand quietly beside their loved one for days, forming a circle around the remains and touching them with their trunks. A herd of elephants will even carry the bones and tusks of their dead comrade for many miles over many days.

Based on a recent sighting in the South Atlantic Ocean, some whales seem to fall in love. Researchers observed two whales embracing and rubbing each other with their flippers after they had mated. When they finally swam away, they continued to touch each other. "They were two peas in a pod," remarked one researcher upon witnessing the whales' obvious affection for one another.

Even dying of a broken heart is possible in the animal world, it seems. Primatologist Jane Goodall, who spent much of her life observing animals in the wild, tells the story of a 50-year-old female chimpanzee who died of old age. Holding her hand and nudging her occasionally, her eight-year-old son refused to leave her side after his mother had died. Whimpering and moaning, he gradually stopped eating and withdrew from the troop. After mourning more than three weeks, he also died. Goodall concluded he died of grief.

Animals show joy and pleasure in the daily lives, too. A happy dog wags its tail and jumps for joy. Cats purr when they're content. And who has not observed seals and dolphins in a playful mood? After a long absence from each other, even elephants greet their friends by flapping their ears, spinning in circles, and trumpeting their reunion.

New research on the brain is also providing more evidence that humans and animals share similar emotions. For example, neuroscientists have discovered that humans and animals share a common brain part called the amygdala. Stimulating this part of the brain creates intense fear. Rats and humans lose their sense of fear when their amygdala is damaged. This fact suggests that humans and rats are wired in a similar way.

Another chemical, oxytocin, controls bonding in both humans and animals. Studies have shown that this hormone plays a key role in maternal bonding and sexual activity in human beings. Research indicates that this hormone governs mating and bonding in some animals as well. When scientists block the flow of oxytocin in female voles, for instance, these mouse-like creatures stop looking for a mate. When this hormone is flowing normally in the voles, they choose a mate in one day.

Further observations and research may well prove that humans and some animals share similar feelings, emotions, brain structures and chemistry.

58. Which of the following is the best way to paraphrase paragraph three? I.C.5, I.C.G11
    A. Animals in cages tell scientists more about animal behavior than observations in the wild.
    B. Despite a growing trend to observation in the wild, researchers are undecided about its benefits.
    C. Scientists often observe animals in their natural habitat, recording their movements and habits through observation and photography, often for use in documentaries and magazine articles.
    D. Watching animals in their native habitat makes for great television.

59. The phrase "two peas in a pod" in paragraph five is an example of I.D.4, I.D.5, I.D.10
    A. imagery.
    B. a metaphor.
    C. alliteration.
    D. a simile.

60. How does the author establish the credibility of Jane Goodall in the passage? I.C.8
    A. He identifies her as a primatologist.
    B. He describes the great work she has done.
    C. He relates the story of the chimpanzee.
    D. He explains that she observed animals for much of her lifetime.

61. Which of the following is most likely the meaning of primatologist? I.B.2, I.B.G6
    A. a scientist who studies chimpanzees or members of the ape family
    B. someone who studies animals in their native habitat
    C. a researcher who observes rituals between animals and their offspring
    D. someone who observes and records animal behavior

62. The word "trumpeting" is an example of the author using _____ to better describe a scene. I.D.4, I.D.5, I.D.10
    A. imagery
    B. a metaphor
    C. personification
    D. a simile

Practice Test One

63. In this passage, the word "trumpeting" most likely means? I.B.2, I.B.G6

    A. making sounds to alert others
    B. roaring to show approval
    C. trying to show authority
    D. a noise to ward off predators

64. Which of the following sentences best summarizes the passage? I.C.5, I.C.G11

    A. In fact, animals and humans share not only common feelings, but also a similar brain.
    B. New research on the brain is providing evidence that humans and animals share emotions.
    C. It has been suggested that humans and rats are wired in a similar way.
    D. Animals and humans share many common characteristics; further research may prove that they share similar feelings, emotions, brain structures and chemistry.

65. Why do you think scientists during Darwin's time did not believe his assertions? I.C.7

    A. Darwin was known to fabricate his theories.
    B. Scientists at the time were ignorant of most animal behavior.
    C. His theories were too different from what scientists already believed to be true.
    D. Darwin's other theories were so outlandish, his credibility had been damaged.

Usually a man writes a love poem to his wife or girlfriend. Read the following unusual love poem from a wife to a husband. Then answer questions 66 – 68.

"To My Dear and Loving Husband"
by Anne Bradstreet

1 If ever two were one, then surely we.
   If ever man were loved by wife, then thee;
   If ever wife was happy in a man,
   Compare with me, ye women, if you can.
5 I prize thy love more than whole mines of gold
   Or all the riches that the East doth hold.
   My love is such that rivers cannot quench,
   Nor ought but love from thee, give recompense.
   Thy love is such I can no way repay,
10 The heavens reward thee manifold, I pray.
   Then while we live, in love let's so persevere
   That when we live no more, we may live ever.

66. What is the symbolism in lines 5 – 6? I.D.5
    A. Her husband's love is compared to mines of gold and Eastern riches.
    B. Bradstreet believes the love she has for her husband is more expensive that gold or riches.
    C. Bradstreet wants more love from her husband.
    D. Bradstreet believes her husband wants more love from her than she could possibly pay.

67. Which best describes the author's tone in lines 1 and 2? I.D.6, I.D.10
    A. angry
    B. heartfelt
    C. joking
    D. mournful

68. Why did Anne Bradstreet write from the first person point of view? I.D.4
    A. Bradstreet wanted to speak about an ideal relationship which did not exist for her personally.
    B. She felt she needed to write as she imagined her husband would write to her.
    C. She wanted to marry an idealized version of love.
    D. She wanted to write about love from her own personal experience.

Read the poem about the Statue of Liberty. Then answer questions 69 – 72.

The New Colossus

1 Not like the brazen giant of Greek fame,
2 With conquering limbs astride from land to land;
3 Here at our sea-washed, sunset gates shall stand
4 A mighty woman with a torch whose flame
5 Is the imprisoned lightning, and her name
6 Mother of Exiles. From her beacon-hand
7 Glows world-wide welcome; her mile eyes command
8 The air-bridged harbor that twin cities frame
9 "Keep, ancient lands, your storied pomp!: cries she
10 With silent lips. "Give me your tired, your poor
11 Your huddled masses yearning to breathe free,
12 The <u>wretched refuse of your teeming shore,</u>
13 Send these, the homeless, tempest-tossed to me,
14 I lift my lamp beside the golden door!"

-Emma Lazarus, 1883, a poem about the Statue of Liberty

69. How does the symbolism in lines 13–14 affect the message of this poem?

   A. It inspired the belief that the United States is a land where gold is plentiful.
   B. The ending with an uplifting light and golden passage leaves the reader with hope.
   C. It leaves the reader to understand that the ships will not crash in the harbor.
   D. The ending creates unfounded fears of death and an afterlife in America.

70. In line 12 the *wretched refuse of your teeming shore* means

   A. the garbage from your beaches.
   B. people who desperately search for a better life.
   C. the excess seafood that is spoiled and left on the beach.
   D. the toxic waste that flows from the rivers into the sea.

71. What aspect of this poem reflects a common assumption in contemporary culture in the United States?

   A. The poem suggests that the United States is a land of opportunity for the oppressed of other nations.
   B. The poem explores the common belief that the United States culture encourages other nations to keep their traditions if they favor the rich.
   C. The poem describes how the United States uses a statue to symbolize itself as a country which guards itself from outsiders.
   D. The poem explores the strong belief in the United States to destroy any established order and create a truly classless society.

72. According to the poem, what is the most significant difference between the "New Colossus" and the old one?

   A. One is bigger than the other.
   B. One is Greek; the other is French.
   C. One symbolizes oppression and the other freedom.
   D. One is of ancient times, and the other is of modern times.

End of Section 3. Check your work.

MCA-II/GRAD Reading Test
Practice Test Two

The purpose of this test is to measure your progress in reading comprehension skills. This test is based on the Minnesota standards for English and Language Arts and adheres to the sample question format provided by the Minnesota Department of Education.

General Directions:

1. Read all directions carefully.

2. Read each question or sample. Then choose the best answer.

3. Choose only one answer for each question. If you change an answer, be sure to erase your original answer completely.

4. After taking the test, you or your instructor should score it using the Answer Key that accompanies this book.

> From slavery to piracy to cannibals, Galveston has seen it all. Read this article about a city with a past. Then answer questions 1 – 9.

Galveston

The island of Galveston has a long and colorful history. Galveston is a barrier island off the coast of Texas in the Gulf of Mexico. Its history is cluttered with shipwrecks, buccaneers, outlaws, cannibals, and explorers. Some of the remarkable characters that have shaped the history of Galveston include the Karankawa Indians, the Spanish explorer Alvar Nunez Cabeza de Vaca, and the swashbuckling pirate Jean Lafitte.

In the 1400s and 1500s, the island was inhabited by natives known as the Karankawa Indians. In Texas, the Karankawa were known for being cannibals. This label, however, is sometimes misunderstood. The Karankawa were a tribe of hunters and fishers. They lived in the coastal area of Texas between Galveston and Corpus Christi. They fished in the shallow waters along the coast and used longbows to hunt. The longbows were as long as the Indians were tall. And the Karankawa were a tall people: many were over six feet tall.

It is true that the Karankawa were known to practice cannibalism, but it was not for food. It was for superstitious reasons. The Indians believed it was a way of gaining power over enemies. A few eyewitness records of the Karankawa still exist. One account, by Alice Oliver, describes the Indians as a tolerant people who taught her some of their language when she was a child in the 1830s.

The Karankawa may have gotten their fierce reputation from the Europeans. These Europeans often tried to kidnap Indians and sell them as slaves. In those days, the slave trade was a large part of the world economy. In their travels as part of this economy, Europeans spread diseases which killed thousands of Indians. They also fought the Indians for their lands. For these reasons, the Karakawa, like many natives, were unfriendly towards Europeans.

One European landed on the island of the Karankawa without a slave ship or any other form of threat to the natives. His name was Alvar Nunez Cabeza de Vaca. He came to the island as a starving survivor of a shipwreck. De Vaca was a famous Spanish explorer. He had sailed from Spain and was out to claim territory for his king and his country. While he was in the Gulf of Mexico 1528, a hurricane washed him onto the shores of Galveston Island.

The Karakawa Indians were generally friendly to de Vaca. He lived among them for four years, treating them with due respect. De Vaca learned about the Indians' way of life. He was able to understand them and communicate with them.

In 1532, de Vaca left Galveston Island and traveled across Texas in search of his fellow countrymen. He found some of them, but the meeting was a shock for his fellow Spaniards. After four years with the Indians, de Vaca looked more like an Indian than a Spaniard. De Vaca wrote of this meeting that the Spaniards he met "just stood staring for a long time."

De Vaca was a writer. In his day, many explorers of the "New World" kept written records of their travels. De Vaca's writings were important for two reasons. First, even today they give us a close-up look at the land and people of North America in the 1500s. The second reason is even more unique to de Vaca. At the time when de Vaca was exploring, European nations were trying to claim as much of the world as they could. Any native people that they met in their quest were considered to be obstacles. These natives were most often conquered, murdered, or enslaved. De Vaca, however, had come to know the natives as people. He was appalled by the way Europeans treated the natives. He wanted to let people know about these injustices.

When de Vaca returned to Spain in 1537, he wrote in graphic detail of the abuses that Europeans were inflicting upon the native people of the New World. His written accounts of cruelty, torture, and murder make a shocking story. They enlightened many Europeans who then called for a better policy towards the natives. Alvar Nunez Cabeaza de Vaca was one of America's first human rights champions.

About 300 years after de Vaca left Galveston, the island was adopted by another colorful personality. In 1817, the "gentleman pirate," Jean Lafitte, came out of no certain origin and arrived on the island. He hoped to set up a base of operation in the Gulf and continue his work of smuggling illegal cargo.

At the time, Texas was fighting for independence from Spain. Jean Lafitte did not become involved in the war. He considered himself a businessman, not a soldier. His shipmates were mostly criminals, and his business often involved illegal activity; but Lafitte's manner was, for the most part, refined. Many people found the handsome "privateer" intriguing. Lafitte set up a small village on the island. There were huts for his pirate friends, places of business, and a red brick mansion built for himself. There was a lot of pirated wealth in the village and constant parties as well.

The island people welcomed Lafitte because of the money, but were uneasy about his friends. On a fall night in 1820, a U.S. government ship threatened Lafitte with death if he did not leave the island. The next morning, the village was in ashes and Lafitte and his band had vanished. There is no record of what happened to Lafitte after that time, but he probably did not retire as a boot maker somewhere. Chances are that he joined the many nameless high seas buccaneers that prowled the Gulf of Mexico in the 1800s.

History still recalls the name of Jean Lafitte, as well as de Vaca and the Karankawa Indians as some of the more captivating chapters in the story of Texas. Today, Galveston's people and cultures are as diverse as ever. This "port of entry" for many immigrants to the United States continues to be a source of stories, legends, and vibrant history.

Practice Test Two

1. The author of this passage is most likely a
   A. historian.
   B. philosopher.
   C. tourist.
   D. pirate.

2. As used in paragraph 8, the word *appalled* means
   A. shocked.
   B. dazed.
   C. worried.
   D. confused.

3. Paragraph 9 is important in this selection because it helps the reader understand
   A. when de Vaca returned to Spain.
   B. what type of person de Vaca was.
   C. how de Vaca viewed the native people.
   D. why de Vaca's writing was important.

4. What is the main reason that the native people welcomed Jean Lafitte?
   A. He was a gentleman.
   B. He was handsome.
   C. He was wealthy.
   D. He built a village on the island.

5. The main reason the Karankawa Indians accepted de Vaca was because
   A. he wrote stories about them.
   B. he treated them with respect.
   C. he taught them how to communicate.
   D. he looked more like an Indian than a Spaniard.

6. Which statement from the selection is an opinion?
   A. Galveston is a barrier island off the coast of Texas.
   B. A few eyewitness records of the Karankawa still exist.
   C. Jean Lafitte had a colorful personality.
   D. De Vaca traveled across Texas in 1532.

7. Paragraphs 11 and 12 of this selection are mainly about
   A. the destruction of the village.
   B. the conflict between Texas and Spain.
   C. the life of Jean Lafitte, pirate and gentleman.
   D. a group of criminals who settle in Galveston.

8. The author probably wrote this selection to
   A. persuade the reader to visit Galveston.
   B. describe the early explorers of Galveston.
   C. provide background information about Galveston.
   D. entertain the reader with a story about the people of modern-day Galveston.

9. From the information about the Karankawa provided in this selection, the reader can infer that
   A. they were hostile toward newcomers.
   B. they were distrustful of the Europeans.
   C. they were involved in illegal activities.
   D. they were unable to communicate with others.

When you listen to the weather report, you hear about wind. Read the following passage about how the wind blows. Then answer questions 10 – 16.

Prevailing Winds

The globe can be divided into six belts by latitude. In each of these belts, there are winds that blow most often. These are called the "prevailing winds" of that latitudinal belt. These belts are marked by their distance — in degrees — from the equator. For instance, one belt extends from the equator — which is 0 degrees — to 30 degrees north of the equator.

If you look closely at the diagram of the bands of the prevailing winds around the world, you will see a pattern in the directions these winds take. If you start at the equator and go north, you will notice that in the first belt, the winds blow from upper right to lower left (or, from northeast to southwest). In the next band to the north, the winds reverse, blowing from southwest to northeast. In the top band, which includes the North Pole, the winds again reverse direction.

Now, if you look at the wind belts from the equator going south, the same reversals occur. This time, however, they start with winds that blow from southeast to northwest.

Winds are named for the direction from which they come. For instance, if winds are called northeastern winds, or "north easterlies," it means that the winds are traveling *from* the northeast.

From the diagram below, you will notice that between 30° north and 30° south of the equator, the prevailing winds are called trade winds. Above the equator they are the Northeast Trade Winds, and below the equator they are the Southeast Trade Winds. If you understand what causes the trade winds, you will understand the basic causes of all prevailing winds.

As has been stated, the trade winds occur near the equator. As the hottest latitude on earth, the equator produces very warm air. Warm air rises and expands. At the equator, the air rises and expands, moving away from the equator. As it flows away from the equator, it cools down. Cool air sinks. Now we have a warm area near the equator, from which the air is leaving, and a cool area about 30° away from the equator, in which there is a lot of cool air. Air always moves from cool to warm. That is why the cooled winds blow back to the equator from the cooler zones around 30° north and 30° south of the equator. This movement of air is the trade winds.

Why are the arrows in the diagram slanted? Why do the Northeast Trades travel back to the equator from the northeast, moving southwest, and not directly due south to the equator? The reason why the prevailing winds do not travel directly north and south is because they are not only influenced by the heat and coolness of the planet, they are also influenced by the motion of the planet. The earth is spinning around, and this makes the winds move in diagonal directions of the compass. The force that the earth's spinning has on prevailing winds is called the *Coriolis* force.

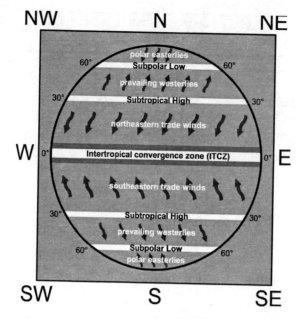

10. The prefix *inter-* in the word <u>intertropical</u> in the diagram makes the word mean
    A. within the tropics.
    B. above the tropics.
    C. below the tropics.
    D. between the tropics.

11. In which direction do the trade winds south of the equator blow?
    A. northeast to southwest
    B. northwest to southeast
    C. southeast to northwest
    D. southwest to northeast

12. Winds that blow from the southeast to the northwest are known as
    A. easterlies.
    B. westerlies.
    C. South Eastern Trade Winds.
    D. north westerlies.

13. The author presents the ideas in this selection mainly by
    A. generalizing and providing explanations.
    B. developing an idea by presenting stories.
    C. using a series of events to illustrate a point.
    D. presenting an idea with personal opinions.

14. What happens as air moves away from the equator?
    A. It rises.
    B. It cools.
    C. It expands.
    D. It warms.

15. Which statement from the selection best supports the main idea?
    A. "The trade winds occur near the equator."
    B. "The equator is the hottest latitude on Earth."
    C. "Winds are influenced by the motion of the planet."
    D. "Winds blow from southeast to northwest."

16. The tone of this piece is best described as
    A. giddy with joy.
    B. scholarly and confident.
    C. confusing and weak.
    D. biased and arrogant.

Did you know there is special training available for rescuing people lost in wilderness areas? Read about one family involved in rescues. Then answer questions 17 – 28.

Rescuing Others a Way of Life for Upper Sierra Woman

Upper Sierra resident Brenda Harrison, along with her dog, her son and her friend were on vacation last week when she was summoned to find two boys lost in the woods.

They were all staying at Joan Heintzelmann's summer home in Mountain Pass, Torro County, when Brenda received a call that two boys were lost in Mount Eden State Forest.

Harrison, along with her dog, Hunter, is a member of the United States Volunteer Wilderness Team. "We're on the State Task Force One, part of the state police office of emergency management," she said. "It's a state disaster team. What we do is train for wilderness and river areas."

Brenda is qualified to go out as a ground team member and is in the process of training Hunter, a German shepherd that weighs almost 100 pounds.

"We packed up and left after I got paged about 10:45 p.m.," she said. "Actually, the park is on Route 42 so it was on our way home. We got there around midnight, talked to the police and asked what they had done so far."

Harrison was told that searchers had driven around and taken tracking dogs into the woods to look for the 9- and 12-year-old boys who had been playing a modern day hide-and-seek game. They were reported missing around 7:30 p.m., she said.

Before going into the woods, the team of three, plus Hunter, picked up another member, a local man who knew his way around the woods.

"When you're out in the middle of the woods, you want to know that everyone on your team has training. Joan and I trust each other, and I trust my son's knowledge," she said.

Steven, 14, a Murdock High School freshman, went to ground search rescue school at Fort Irwin last spring, a program run by the Civil Air Patrol. "It's a great organization for kids," she said.

"Steven knows how to use a compass and a map," Harrison said. "He was my communication person. He was talking to the state police the whole time. He also helped carry the 9-year-old boy out of the woods."

"Before going in, we studied a map of the park," Steven said. "We were walking through massive briars and trees. It was 35 degrees that night. We found them about 1:30 a.m," Brenda added. "They weren't dressed for such a drop in temperature, but they were fine, just very cold," she said. "I don't think they will go that deep in the woods any more."

Harrison is modest about the rescues she has been involved in. "It's just what I do," she said. "The next day I go back to my normal life. It's a way of putting my talent to good use."

Her normal life is a full one. Besides Steven, she has two daughters and is a full-time student studying for a dual degree in education and special education.

She takes her dogs to class with her. "A lot of time I train my dogs after class," she said. "We train all over Nevada. Training has to be continuous in order to reinforce what they've been taught. Hunter thinks it's a game. He loves to go to work."

She and her husband are both members of the Upper Sierra First Aid Squad, which sometimes calls her to aid them with her dogs. She said she and her team members are going back to Mount Eden State Forest to look around. "I need to see it during the daytime. I remember how massive the woods looked."

17. According to the information in the article, why would the search and rescue volunteers work in teams?   I.D.7 I.D.5 I.C.G12
    A. so the dogs don't fight with each other
    B. because each individual has special knowledge to contribute to the team
    C. to keep the dogs calm and focused on the rescue
    D. to fulfill the safety requirements of the police who are involved

18. How many people were on the volunteer rescue team that found the boys in the woods?   I.C.5 I.C.G11
    A. 4 + Hunter
    B. 3 + Hunter
    C. 2 + Hunter
    D. 1 + Hunter

19. How long were the boys missing?   I.C.5 I.C.G11
    A. 1 1/2 hours
    B. 3 hours
    C. 5 hours
    D. 6 hours

20. What special qualification did Steven Harrison have for working on the rescue team?   I.C.5 I.C.G11
    A. how to communicate and coordinate with the state police
    B. able to help carry the boys out of the woods
    C. attended ground search and rescue school
    D. was a Murdock High School freshman

21. Based on the passage, we can infer that police departments appreciate the work of volunteers because   I.C.7 I.C.G12
    A. officers can have more time off with their families.
    B. search parties have more people, but for the same price.
    C. volunteer dogs are the only ones they can depend on.
    D. they save the police the cost of a rescue dog.

22. In this article, the writer's main focus is to
    A. explain the importance of first aid.
    B. analyze the special skills that a rescue dog provides.
    C. describe the emotional response to finding missing children.
    D. profile a member of the Wilderness Rescue Team.

23. Read the following excerpt from the passage:

    "... a local man who knew his way around the woods."

    Using context clues, we can infer that the man
    A. could find a way to go around the woods to get to the other side.
    B. loved the animals in the woods.
    C. would not get lost in the woods.
    D. was very educated in forestry.

24. Which statement is *best* supported by the information in the article?
    A. Trained dogs are skilled at search and rescue.
    B. Trust and teamwork are important in search and rescues.
    C. Children need survival skills before venturing into the forest.
    D. Mount Eden State Park is a dangerous place.

25. Which of the following lines from the article *best* describes Brenda Harrison's personality?
    A. Her normal life is a full one.
    B. She takes her dogs to class with her.
    C. She is modest about the rescues she has been involved in.
    D. She and her dog, Hunter, are members of the United States Wilderness Team.

26. Most of the events described in the article take place
    A. at Brenda Harris's home.
    B. in the woods in Nevada.
    C. in the classroom.
    D. at the University Brenda Harris attends.

27. The tone of this piece is best described as
    A. impassioned.
    B. indifferent.
    C. hopeful.
    D. objective.

28. Write a summary in two or three sentences for the passage.

End of Section 1. Check your work.

> The main way our government gets money to run is through taxes. Read the following article about taxes and other ways our government gets revenue (money). Then answer questions 29 – 36

A Brief Overview of Government Revenue

The United States government is known worldwide as a leader in managing the economic and political affairs of the nation and for advising its international neighbors. To explain in detail is to use the example of how the federal government has worked out situations affecting trade of goods between the states and also abroad. In addition, its domestic policies in relation to revenue generating are worth noting. Domestically, we trade goods that are manufactured or grown in the country between all the states with relatively little problem; but when we ship those goods overseas, it becomes complicated. The process itself is called *exporting* and how well it works depends upon certain factors, such as the quantity of goods shipped, what the goods are being transported, where the goods are being shipped to, et cetera. The United States also takes in goods from other countries through a process called *importing*. The federal government monitors the nation's inventory as well as its economic flow of goods shipped out or taken in through a system called *balance of trade*. It is important to be mindful of how many goods are exported versus how many are imported because if imported goods are worth more in value than exported goods a *trade deficit* results. A trade deficit can be harmful to various sectors of the nation's economy if it continues for a prolonged period of time.

In negotiating agreements, policies, or treaties with other countries, it is important to have diplomatic recognition. This means it is not only essential to be aware of cultural differences between various nations but to also be skillful at managing the relations between all of the different countries involved. Many factors are involved in doing business with people of foreign lands, since all have different ideologies and practices. It requires patience and education to understand these differences and to make sound international trade agreements.

Domestic policies also encompass a wide range of issues and matters from government revenue to the amendments within the Constitution. To begin with, there are multiple sources of government revenue. One of the largest sources of money for the federal government is taxes on citizens. There are different taxation systems. One type is called a *proportional* tax. This means that the same percentage comes from every citizen, regardless of variation in income earned by each individual. A second type of tax is called the *progressive* tax. This tax is based upon the amount of income an individual earns. The progressive tax system uses a sliding scale to determine how much a person pays. *Regressive* tax is a tax placed upon products, materials, services, et cetera and does not factor in the amount of income a person earns. An example of this type of tax is sales tax. In this example, people choose how much they pay by making the decision to buy or not to buy items.

There are also specific types of tax the federal government collects from an individual's income. The main one is the federal income tax, which currently uses the progressive tax system mentioned earlier. There is also FICA, which contributes to the federal social security system. Another way the federal government receives taxes is by *tariffs*. In the past, the government taxed imported and exported goods (even between states), but now taxes imports exclusively.

Other sources of government revenue would include generating money from fines, licenses, user fees and from a person borrowing money such as on a loan from a bank. Fines would include a penalty an individual would pay to the Internal Revenue Service if federal income taxes were underpaid or if a company violated Environmental Protection Agency regulations. Revenue that comes from borrowing passes from the lending rate the government gives to banks for "borrowing" the money and then to the individuals borrowing the money from the banks. This rate passes from the government, to the banks and then to individuals, in the form of an *interest rate*.

29. According to the passage, government tariffs would include
    A. items shipped from New York to Baltimore.
    B. items shipped from New Orleans to Tokyo.
    C. Anything the United States manufactures.
    D. items brought in from Thailand.

30. The use of the word "complicated" in Paragraph 1 mostly likely means
    A. complex and technical.
    B. open to interpretation.
    C. hard to understand.
    D. not worth bothering about.

31. It can be inferred from the last paragraph about the Environmental Protection Agency that fines are paid by
    A. individuals who litter.
    B. private businesses that pollute illegally.
    C. other nations who do business in the United States.
    D. people that cheat on their income taxes.

32. The word "exclusively" in paragraph 4 most nearly means
    A. harshly.
    B. without regard to right or wrong.
    C. only.
    D. since the beginning of the nation.

33. The narrator explains the different types of earning opportunities for governments in order to
    A. show how the United States government gets money to run its programs.
    B. explain how the government exploits other nations.
    C. argue how and why taxes are unfair to working people.
    D. show how different nations earn money.

34. In paragraph three, the author explains different kinds of taxes by
    A. classifying each one and favoring the progressive tax.
    B. listing the definition and legal regulation.
    C. naming each and explaining how it works.
    D. providing the history of each tax.

35. The first paragraph of the passage serves to
   A. explain why the United States is different than other nations.
   B. introduce the reader to the complexities of international relations and revenue generation.
   C. claim the United States should do more for other nations.
   D. argue that taxes are wrong.

36. It can be inferred from the passage that international relations
   A. are based on low tariffs alone.
   B. involve diplomacy, negotiation, and a strong military.
   C. explain that unfair trade balances can lead to war.
   D. often hinge on fair and balanced trade with other nations.

> Most of us take United States Citizenship for granted. Read the following interviews with those who consider it a great privilege. Then answer questions 37 – 42

Becoming An American At Last

There were no doubts or second guesses. Salahuddin Jamil knew he wanted to be an American. "I had experience in Europe and America and my home, a third world country," said Jamil, a Bangladesh native. "I have seen three sides, and there is no greater country than the USA."

Each year, Congress decides how many immigrants may enter America, and the number permitted has risen steadily. So has the rate of assimilation. A recent Census data report by the National Immigration Forum indicates that more than two-thirds of U.S. immigrants speak English fluently within 10 years. It is largely this group that accounts for the increase in applicants to the Immigration and Naturalization Service seeking to become American citizens.

But assimilation takes more than desire. The naturalization process is lengthy – taking as long as two years between the application and the citizenship oath. It includes background checks, fingerprinting and tests – written and oral – with questions about history and politics.

Oksana Gerasimenko, a 31-year-old Ukrainian refugee, was among thousands who took the oath in 1999. "The U.S. has always been a country of freedom, not England, not Japan or anywhere else," said Gerasimenko. "That's what I thought as I was growing up. I got that from the movies and the books."

When Gerasimenko moved to Atlanta, she knew of about two dozen Russians in her area. But she's noticed that the population has increased lately to almost 30,000. "Now, I hear Russian [spoken] in the grocery store, and anywhere I go," she said.

Gerasimenko's husband decided the couple would move to Atlanta in 1991 because he felt the Olympics would create business opportunities. He now owns his own construction company and is among the thousands waiting to take the Oath of Allegiance.

Jamil, 41, bought his first restaurant two years after becoming a citizen. He worked as a cook in an Indian restaurant and saved the bulk of his earnings. Then he made an agreement with a soon-to-be-retiring couple that if he successfully ran their delicatessen, he could buy it from them at a bargain.

He now owns a successful sandwich shop in downtown Atlanta across the street from the Immigration and Naturalization Service offices.

"John Kennedy's Peace Corps gave me respect for the USA," said Jamil. "Then I read about Martin Luther King, and that is why I had to come here [to Atlanta]. We can eat together and stand together and work together because of him."

Jamil speaks proudly of the franchises he has owned and sold, and the addition of hand-dipped ice cream to his store. He becomes passionate and forceful when he talks about the homeless drunk he hired and sent to a rehabilitation center.

"We must do for the weaker person. All they need is love," he preaches with a clenched fist. "No one here is better than anyone else." It's part of the reason Jamil and several of his countrymen started the Bangladesh Association, of which he is now a commissioner. The organization works toward civic and social goals.

As Gerasimenko waited in line to relinquish her green card and sign her citizen certificate, she said she was nervous. "For the test, you can prepare," she said. "This you can't."

Holding the flag just beneath the oath she would soon repeat, Gerasimenko looked solemn as a recording of the Star-Spangled Banner played overhead. She stood between a woman from the Ukraine and a man from Vietnam and repeated the Oath of Allegiance. Her vow to remain true to America was easy to make, she said.

"I didn't have any place where I was a citizen," Gerasimenko said. "So this was not hard."

37. As used in paragraph three, "assimilation" can be inferred to mean
    A. adopting the customs and traditions of a foreign place.
    B. converting into sound.
    C. learning to speak a language.
    D. the rate of immigration.

38. All of the following are facts taken from the passage EXCEPT—
    A. Immigrants don't work as hard as American citizens.
    B. More immigrants want to become U.S. citizens.
    C. The rate of immigration to America grows every year.
    D. Becoming an American citizen is a complicated process.

39. The theme of this passage can best be expressed as
    A. the United States is a country of freedom.
    B. immigrants really enjoy all of their experiences in the United States.
    C. immigrants are becoming responsible citizens in the United States.
    D. Kosovo refugees are making new homes in the United States.

40. In line 167, "fluently" most nearly means
    A. with great skill.
    B. with great difficulty.
    C. only with others of the speaker's own country of origin.
    D. only to family members.

Practice Test Two

41. The story of how Jamil came to own a delicatessen in Atlanta serves to   I.C.6 I.C.9 I.C.8

    A. verify his claims about life in America.
    B. act as a warning to other immigrants.
    C. explain that Atlanta is a good city for immigrants.
    D. dramatize a successful immigrant experience.

42. According to the passage, Jamil was inspired to come to the United States after   I.C.5

    A. fleeing a war in his own country.
    B. growing dissatisfied with life in Europe.
    C. learning about the Olympics.
    D. observing the Peace Corps and learning about Martin Luther King.

> To stay with the North or side with the South, these were tough questions for Maryland citizens at the beginning of the Civil War. Read how Abraham Lincoln helped them decide. Then answer questions 43 – 52.

Maryland in the Civil War

1   By the beginning of the 1860s, the secessionist and abolitionist crises throughout the states and territories had come to a boiling point. A number of causes urged voices in the South towards secession, but at their heart was the issue of slavery. Despite his assertions to the contrary, leaders in the Southern states believed Abraham Lincoln would challenge the two-century old practice, which would mean a crippling loss of cheap (albeit unethically-gained) labor. For Lincoln, maintaining the union of states was more important than any other concern, including his personal dislike of the institution of slavery. The Southern states were also angered by the fact that Northern states did not always comply with the Fugitive Slave Clause of the Constitution, which promised that runaway slaves would be returned to their
10   owners and were not free simply because they moved to a free state. The *Dred Scott v Sanford* decision, in which the Supreme Court stated that former slaves in free states did not have legal standing because they were not citizens, dealt a powerful blow to abolitionists but did not end the controversy. Northerners were unhappy with the decision because it overturned the free-soil movement that attempted to build a slavery free land in the West. Other issues played a role as well. Northerners wanted tariffs on manufactured goods to protect their industries, but Southerners opposed tariffs because they wanted cheaper manufactured goods and did not want to endanger the lucrative trade relations they had with Britain.

      As tensions rose on both sides of the Mason-Dixon line, the country armed itself and braced for what almost everyone understood to be inevitable. On December 20, 1860, South
20   Carolina decided to secede from the Union. Six other states followed suit by February 1861. Four more would withdraw from the Union when the war began at Bull Run, Virginia on July 21, 1861. This would bring the total number of secessionist states to eleven, enough to outweigh the Union in size if not population and industrial development; enough to become a credible nation in its own right if not checked, one that would almost certainly side against a reduced United States and with its trade-hungry European rivals.

      As the battle lines were drawn, Maryland was split between the North and the South. Maryland was a slave state but did not rely on plantation agriculture to nearly the degree the Deep South did. Evidence taken from the memoirs of Maryland slaves, most notably the great abolitionist and orator Frederick Douglass, suggests that, compared to their brethren farther
30   south, slaves in relatively industrialized Maryland were treated more humanely and

given educations and comforts uncommon among slaves in the Cotton Belt of the Deep South. Nevertheless, Maryland bore the weight of both sides of the crisis: owning slaves and wanting to abolish slavery once and for all. Economically, Maryland's wealth was hinged upon its port city, Baltimore; the state was beholden to the industrial centers of the north — most notably Pennsylvania, New York, and Massachusetts — for export of goods, and trembled before the possibility of separation from its northern client-states. Continued slavery and secession would mean losing the markets of those states' cities, devastating its economy.

Four slave states remained on the side of the Union (Missouri, Kentucky, Delaware, and Maryland), but southern sympathizers were common in these states and routinely pressured
40 their governments for secession. To the Union, the secession of Maryland represented a threat to Union border security. If Maryland joined the Confederacy, Washington, DC would find itself surrounded by Confederate territory. Concerned that Confederate sympathizers might succeed in swaying Maryland into the Confederate camp, President Lincoln declared martial law in Maryland and suspended the right of *habeas corpus*, which guarantees a person cannot be imprisoned without being brought before a judge. The president then jailed the strongest supporters of the Confederacy. As a result, Maryland backed down and its legislature voted to remain in the Union. The suspension would not be lifted until the end of the Civil War, four years later.

43. In line 7, "institution" most nearly means
    A. formal place of learning.
    B. mental asylum.
    C. correctional facility.
    D. long-standing practice.

44. The author indicates that the end of the free soil movement and tariffs were
    A. reasons the South wanted to secede from the North.
    B. important reasons for Europe to boycott the South.
    C. reasons the North wanted a war with the South.
    D. reasons the North wanted to avoid a war.

45. The term "tensions" in paragraph 2 refers to the
    A. feeling of unease had by one side of the secessionist debate.
    B. heated disagreements between North and South.
    C. strain to fit everyone in a limited land space.
    D. potential consequences of secession.

46. The author cites the Dred Scott Decision in the first paragraph in order to
    A. describe a particular event in the months leading to the war.
    B. show how laws about slaves were already in place.
    C. demonstrate how the South ignored laws passed in the North.
    D. show the effects of the abolitionist cause.

47. The third paragraph presents a(n)   I.C.5   I.C.7   I.C.G12
    A. overview of conditions in Maryland.
    B. list of reasons for Maryland to secede.
    C. list of sources.
    D. introductory aside.

48. In paragraph 3, the phrase "relatively industrialized" indicates that   I.B.2   I.B.G6   I.D.14
    A. there were few industries in Maryland.
    B. there were more industries in the South than Maryland.
    C. the industries of North and South were related.
    D. Maryland was more industrialized than the Southern states.

49. The phrase "hinged upon... Baltimore," within the context of paragraphs three and four, indicates that   I.B.2   I.B.G6   I.D.14
    A. Maryland's economy depended on the prosperity of Baltimore.
    B. Baltimore was the only opening for goods into the South.
    C. the North would occupy Baltimore if war broke out.
    D. the South wanted to invade Baltimore.

50. The attitudes towards secession and abolitionism in this article can be described as   I.D.6   I.D.10   I.C.6   I.C.9
    A. uninterested and uncaring.
    B. enthused and involved.
    C. open-minded and hopeful.
    D. objective and interested.

51. In paragraph four, "the Confederate camp" is understood to mean   I.B.2   I.B.G6
    A. all the states and people in favor of secession.
    B. anyone outside Maryland.
    C. everyone in the South.
    D. only those people in Maryland who wanted slavery.

52. Do you think the author of   I.C.5, I.C.G11, I.C.7, I.C.G12, I.C.8, I.D.4
this passage was objective in reporting conditions in Maryland during the civil war? Why or why not? On a separate sheet of paper, explain your position in a well-structured paragraph.

MCA-II/GRAD Reading Test

For many events in life, a social security card is required. Read over this application for a social security card. Then answer questions 53 – 57.

SOCIAL SECURITY ADMINISTRATION
Application for a Social Security Card

1. NAME TO BE SHOWN ON CARD — First / Full Middle Name / Last
   FULL NAME AT BIRTH IF OTHER THAN ABOVE — First / Full Middle Name / Last
   OTHER NAMES USED
2. MAILING ADDRESS (Do Not Abbreviate) — Street Address, Apt. No., PO Box, Rural Route No. / City / State / Zip Code
3. CITIZENSHIP (Check One) — ☐ U.S. Citizen ☐ Legal Alien Allowed To Work ☐ Legal Alien Not Allowed To Work (See Instructions On Page 1) ☐ Other (See Instructions On Page 1)
4. SEX — ☐ Male ☐ Female
5. RACE/ETHNIC DESCRIPTION (Check One Only - Voluntary) — ☐ Asian, Asian-American or Pacific Islander ☐ Hispanic ☐ Black (Not Hispanic) ☐ North American Indian or Alaskan Native ☐ White (Not Hispanic)
6. DATE OF BIRTH — Month, Day, Year
7. PLACE OF BIRTH (Do Not Abbreviate) — City / State or Foreign Country / FCI
8. A. MOTHER'S MAIDEN NAME — First / Full Middle Name / Last Name At Her Birth
   B. MOTHER'S SOCIAL SECURITY NUMBER — ☐☐☐-☐☐-☐☐☐☐
9. A. FATHER'S NAME — First / Full Middle Name / Last
   B. FATHER'S SOCIAL SECURITY NUMBER — ☐☐☐-☐☐-☐☐☐☐
10. Has the applicant or anyone acting on his/her behalf ever filed for or received a Social Security number card before?
    ☐ Yes (If "yes", answer questions 11-13.) ☐ No (If "no", go on to question 14.) ☐ Don't Know (If "don't know", go on to question 14.)
11. Enter the Social Security number previously assigned to the person listed in item 1. — ☐☐☐-☐☐-☐☐☐☐
12. Enter the name shown on the most recent Social Security card issued for the person listed in item 1. — First / Middle Name / Last
13. Enter any different date of birth if used on an earlier application for a card. — Month, Day, Year
14. TODAY'S DATE — Month, Day, Year
15. DAYTIME PHONE NUMBER — Area Code / Number
16. YOUR SIGNATURE — I declare under penalty of perjury that I have examined all the information on this form, and on any accompanying statements or forms, and it is true and correct to the best of my knowledge.
17. YOUR RELATIONSHIP TO THE PERSON IN ITEM 1 IS: ☐ Self ☐ Natural Or Adoptive Parent ☐ Legal Guardian ☐ Other (Specify)

DO NOT WRITE BELOW THIS LINE (FOR SSA USE ONLY)

EVIDENCE SUBMITTED

SIGNATURE AND TITLE OF EMPLOYEE(S) REVIEWING EVIDENCE AND/OR CONDUCTING INTERVIEW

Form SS-5 (11-2002) EF (11-2002) Destroy Prior Editions   Page 5

53. What is the BEST description of the purpose of this document? I.C.3
    A. to instruct people how to apply for a Social Security number
    B. that the material can be used as a workplace document helping new employees understand Social Security numbers
    C. as an information-gathering government document for an application for a Social Security number
    D. for use as a public document or product

54. Which of the following statements BEST describes the structure of the document? I.C.3
    A. The structure is simple and made for public use so citizens can apply for a Social Security card without going through the government agency.
    B. A smaller heading would help the structure flow evenly and give equal importance to other parts of the document.
    C. Charts would have helped the structure of this document focus on its purpose and audience.
    D. The structure is complicated, but it is well organized by numbering and grouping the material.

55. Which of the following statements BEST describes the document headings? I.C.3
    A. The headings are typical of a government document: the title is largest, and all other headings are equal to each other, except for optional information.
    B. The headings are all the same, and so all sections are similar in importance and applicants must completely fill them out.
    C. The document's headings help display the meaning of the graphics in an organized and structured form.
    D. All the sections have all different type headings since the document's purpose is to make the report easy for government office employees to fill out.

56. The format is BEST described by which of the following statements? I.C.3
    A. The document is informative in format: it has all the information that a person will need to know to apply for Social Security benefits.
    B. The author's purpose in the formatting of the document was to leave space for projected benefits.
    C. The format follows the usual company material, needing to know the employee's name and work area.
    D. The document is information-gathering in format: it asks for a fairly detailed history of the person applying for the Social Security card.

57. The form is best classified as a _____ document I.C.3
    A. public
    B. consumer
    C. private
    D. government

End of Section 2. Check your work.

Section 3

> Modern medicine usually dictates the way women will give birth to their babies, but is this nature's way? Read about the issues in the following article. Then answer questions 58 – 65.

A New Way to Give Birth?

Where is the safest place for a mother to give birth to her baby? You may think a hospital, but the facts show that the mother's own home is just as safe — if not safer — when a trained midwife attends the birth. This may seem like a strange idea, but for millennia women gave birth at home; only in recent centuries have they come to natal term in hospitals. Age-old wisdom and modern science are coming together to show that for a woman with a healthy pregnancy, her own home is the safest birthplace because she is in a familiar environment, surrounded by loved ones and free from standardized hospital policies.

Pregnant women must feel safe and comfortable or they will have interrupted labor. This fact is recognized in the animal world but often neglected in regard to human mothers. For example, when a doe feels labor coming on, she seeks out a secure and protected place to give birth. If she detects a predator approaching, her contractions will stop, so that she can run away and find another safe spot. Once settled in, her labor will resume. In a similar way, many women progress well in their labor at home, but when they arrive at the hospital their labor slows down. Sometimes, the hospital even sends the woman home, where her contractions will resume normal progression. The disruption and anxiety of the trip to the hospital can cause a woman's labor to slow or halt. If she remains in the safety and comfort of her home throughout labor and delivery, her birth experience will probably follow a more natural and rhythmic pattern.

A more important factor in the woman's feeling of safety and comfort is the presence of loving and supportive friends and family. This may seem like a nice extra that has little biological effect on the birthing process, but there are significant differences between births of women with and without help. The comforting presence of a friend or family member leads to diminished desire or need for anesthetic drugs and fewer surgical procedures known in medical terminology as "Cesarean" sections. Many hospitals still place limitations on who can be with the mother while she is giving birth. There is also little space in one hospital room. At home, friends and family are welcome to come and go as the mother chooses, not as the hospital dictates.

The number of people allowed to be with the mother is just one restriction hospital policies place on birthing mothers. Freedom from these restrictive policies is probably the best part of giving birth at home. Hospitals require birthing mothers to eat nothing more than ice chips during labor, so that the mother's stomach is empty in case she needs general anesthesia for an emergency Cesarean. This deliberate starvation causes the mother to lose the energy she needs to deliver the baby, which in turn increases the need for a Cesarean section. Furthermore, hospitals require women to lie down in bed for electronic monitoring of the baby. This electronic vigilance allows fewer nurses to be on staff and provides insurance companies with a record of the baby's health. However, it prevents the mother from walking — a natural way of using motion and gravity to facilitate the birthing process. Finally, hospitals routinely use drugs which severely affect the unborn or newborn child, causing the mother to be less able to enter contractions or use the Lamaze method. These practices can benefit some women, but when they are empirically applied to all women across the board, they can make labor more strenuous. Under the care of a midwife at home, a woman has more flexibility with procedures that aid her unique birth.

For women who are experiencing medical problems in the birth process, hospitals provide the necessary interventions to save lives. However, these interventions are unnecessary for a normal, healthy birth. In this case, a woman can have a wonderful birth experience with loved ones nurturing her in her own home and with a knowledgeable midwife attending to her needs.

58. The use of the term "age old wisdom" refers to   I.B.2 I.B.G6

    A. common knowledge that ordinary people have used for centuries.
    B. the latest gossip.
    C. ignorance and superstition.
    D. previous beliefs about midwifery.

59. In paragraph four, the author describes hospitals as being   I.D.6 I.D.10 I.D.14

    A. mean and uncaring.
    B. sterile and professional.
    C. warm and compassionate.
    D. cold and impersonal.

60. The central contrast between the wilderness in paragraph 2 and the hospital in paragraph three is best described in which terms?   I.D.7

    A. theory versus practice
    B. nature versus modern medicine
    C. imaginary versus reality
    D. expectation versus result

61. The passage serves mainly to   I.C.6 I.C.9 I.D.7 I.D.14

    A. inform readers about the dangers of midwifery.
    B. argue that hospitals are unclean places for childbirth.
    C. discuss the history of childbirth practices.
    D. argue for natural childbirth performed at home.

62. "Cesarean sections" most likely involve   I.B.2 I.B.G6

    A. midwives.
    B. complicated medical procedures.
    C. painful labor techniques.
    D. doctors working with midwives.

63. Which of the following, if true, would undermine the author's argument about the virtues of natural childbirth at home?   I.C.6 I.C.9 I.D.7 I.C.8

    A. Births in hospitals are less susceptible to infection or complications.
    B. More infants die in hospitals than at home.
    C. People go to hospitals to die.
    D. Doctors do not like children.

64. The author mentions contractions and Lamaze in order to   I.C.6, I.C.9, I.D.7, I.C.7, I.C.G12, I.D.4

    A. give examples of natural childbirth techniques.
    B. support the worth of a Cesarean section.
    C. prove her theories about birth in the wild.
    D. discredit doctors.

65. Do you believe expectant mothers would do better to have their children at home, or in hospitals? On a separate sheet of paper, give reasons to defend your position, drawing from both the passage and your own experience.   I.C.8 I.C.G11

A young man worries about his relationship with the woman he loves, and it spills over into his dreams. Read the following poem about a young man's worries. Then answer questions 66 – 68.

"I Dreamt..."

by Suleiman Rustam

1 I dreamt last night that we had parted.
   We strolled no longer through the lane.
   No nightingale its trilling started
   In gardens sear – all life was pain

5 My misery like a cloud above me
   Deprived my soul of laughter's lilt
   I asked, "Why does my love not love me?
   In what, O love, consists my guilt?"
   My tortured thoughts were turbid, sickly.
10 To questions there was no reply.
   How could her fancy change so quickly?
   How could such love abate and die?

   I woke to memories reassuring,
   For yesterday, to my great pride,
15 I heard sweet words of love, ensuring
   That you will never leave my side.

   To die a little, parting seems,
   So let us only part in dreams.

66. Which does the figurative language illustrated in lines 1 – 4 emphasize?   I.D.4
    A. the destruction of war
    B. the pain of separation
    C. the blooming of gardens
    D. the painful tilling of the soul

67. Which of the following best describes the change in the author's tone from the beginning to the end of the poem?   I.D.6, I.D.10
    A. happy to sad
    B. alert to sleepy
    C. anguished to relieved
    D. angry to peaceful

68. Which of the following best describes the author's feelings that are reflected in the poem?   I.D.4
    A. He enjoys sleeping so he can have some time for himself.
    B. He is deeply in love and wants his partner to always feel the same way.
    C. He is not sure that his love will always want to be with him.
    D. He is tired of having recurring nightmares about his love leaving him.

Practice Test Two

> The poet notices a little seedling becoming a plant and compares it to human life. Read the poem. Then answer questions 69 – 72.

The Seedling

Paul Laurence Dunbar

1   As a quiet little seedling
      Lay within its darksome bed,
      To itself it fell a-talking,
      And this is what it said:

5   I am not so very robust,
      But I'll do the best I can;"
      And the seedling from that moment
      Its work of life began.

9   So it pushed a little leaflet
      Up into the light of day,
      To examine the surroundings
      And show the rest the way.

13   The leaflet liked the prospect,
      So it called its brother, Stem;
      Then two other leaflets heard it,
      And quickly followed them.

17   To be sure, the haste and hurry
      Made the seedling sweat and pant;
      But almost before it knew it
      It found itself a plant.

21   The sunshine poured upon it,
      And the clouds they gave a shower;
      And the little plant kept growing
      Till it found itself a flower.

25 Little folks, be like the seedling,
    Always do the best you can'
    Every child must share life's labor
    Just as well as every man.

29 And the sun and shower will help you
    the lonesome, struggling hours,
    Till you raise to light and beauty
    Virtue's fair, unfading flowers.

69. Who is speaking in lines 25 – 32?   I.D.6, I.D.10
    A. the sunshine - third person
    B. the stem - third person
    C. the seedling - third person
    D. the narrator - third person

70. Read the lines from line 18 in the poem.   I.D.4
    What effect does Dunbar want you to get by personifying the seedling sweating and panting?
    A. the freshness of the plants
    B. the struggle of plant growth
    C. that talking to plants helps
    D. not to cut living things

71. What is the theme of the poem?   I.D.6, I.D.10
    A. You can work your way to the top by helping others.
    B. We may not be sturdy and strong, but we can all do our best and share life's labor.
    C. We all begin in the dark and may see the light.
    D. The key to plant growth is sunshine and rain.

72. Read the lines from line 5 in the poem.   I.B.2
    Which dictionary definition of the word *robust* best applies to its use in these lines?
    A. enthusiastic
    B. strong
    C. peppy
    D. zesty

End of Section 3. Check your work.

Appendix
Games and Activities

VOCABULARY (WORD MEANING)

Here are some suggestions for vocabulary games and activities:

1. **Pop Poetry.** Cut out new words, phrases, and accompanying pictures from newspaper and magazine ads, articles, or catalogues. Enlarge the words if they are too small. Make poems or collages by gluing them together on construction paper or poster board. Explain the meanings of the new words and the theme or message of the poem or collage.

2. **Crossword Puzzles.** Locate crossword puzzles in magazines, newspapers, books, or Websites. Working in pairs, students guess at the letters that fit the horizontal and vertical boxes. Looking up some words in the dictionary may be necessary to encourage vocabulary development.

    If you want to learn content-specific vocabulary words, *Crossword Magic* (HLS Duplication, Inc.), a computer software program, takes lists of words and clues and automatically creates a crossword puzzle.

3. **Test Creator Software.** Teachers and students can create their own vocabulary or comprehension tests. This program creates multiple-choice and true/false questions for any subject. Hints and feedback can also be included with the tests. In addition, tests show student scores and track time on tasks. *Test Creator Software* can also be used for classroom presentations and demonstrations with features such as spotlight, magnify, zoom, underline, edit, create sounds, and more. You can download a free demo disk at www.americanbookcompany.com. To place an order, contact American Book Company toll free at 1-888-264-5877.

4. **Shopping Trip.** Students sit in a circle of 5 – 6 persons. In a class, several circles would be formed. The leader starts the chain by saying, "Today, I'm shopping for a short story for dinner. I'll need some ingredients like a plot. A plot is a series of events leading to a climax and resolution." The next student says, "The leader is going to buy a plot, and I am going to buy a character. A character is.. . " Continue around the circle until all of the ingredients are identified and defined. Then the leader can ask, "Who remembers what anyone bought?" Students can volunteer answers such as, "remember that Ben bought some conflicts."

Games and Activities

Variation. For variation, students can shop for types of clothing, parts of a car, house, etc. For example, the leader could say, "I'm dressing up for my poem, so I'm going shopping for a verse. A verse is...."

5. **Word Game.** Each week students bring two words and their definitions to class. Students can choose these words from newspapers, magazines, books, television, etc. Weed out duplicate words. In the following week, students can either take a quiz by defining the words, using them in sentences, or taking a multiple-choice test. Students can also form teams with the teacher asking questions and providing prizes for the best scores.

6. **Semantic Map.** A semantic map is a visual aid that helps you understand how words are related to each other. You write a word or concept in a circle in the center of a page. Then you draw branches out of the center of the circle with related words in circles at the end of these branches. You can work in pairs or in groups. For example, the word collaborate relates to the words cooperate, team, communicate, etc. The root word **super** relates to such words as superior, supersede, superstar, etc.

READING COMPREHENSION

Here are some suggested games and activities:

1. **Prereading.** Before reading an article or story, preview the title, the first paragraph, first sentences of each of the other paragraphs, the last paragraph, and any illustrations. Then write down and discuss what you already know about the article or story. This prereading activity helps improve comprehension.

2. **5 Ws and H.** After you preread or skim a selection, you can increase your comprehension by developing questions based on the 5 Ws and H. The 5 Ws stand for *who, what, when, where,* and *why*. The H refers to *how* something happened or was done. After prereading, you should write down questions about the selection based on the 5 W's and H. After reading the selection, you then answer these questions, rereading the text as needed to confirm your answers.

3. **K-W-L.** With K-W-L, you can improve your comprehension of many types of literature. It involves activities you can do before and after you read. K stands for what you already *know* about the article. W refers to what you *want* to know about the article. L stands for *what* you learned from the article. Discussing and writing down your K-W-L thoughts will help you remember what you read.

4. **Think-Alouds.** With think-alouds, you talk to yourself and ask questions while reading a selection. In this way, you can think about and understand the ideas better. You can also form thoughts and questions and write them down as you are reading. As a result, you are starting to learn and retain what you are reading.

5. **Semantic Map.** In format, a semantic map for comprehension is similar to the one you learned about under Vocabulary (Word Meaning). Since it represents ideas visually, it can help you remember what you have read. To summarize what you have learned in a reading selection, create a semantic map of the main ideas and supporting details in the article. Then use it for recalling and reviewing what you have read.

6. **Split-Half Sheets.** With a split-half sheet, you can review the main ideas and facts in a reading selection. Fold a sheet of paper in half lengthwise. On the left side of the sheet, write the questions you wrote based on the 5 Ws and H and K-W-L. On the right side of the sheet, write the answers to each question in your own words. Review the reading selection to check your answers. Then fold back the answer side of the sheet so you cannot see the answers. Then read the questions only and answer as many you can. Place a check besides the questions you got right, and review the ones you missed until you can answer them correctly.

7. **Make Your Own Tests.** You will read more effectively if you make your own tests. After you complete a reading selection, work in a small group to develop practice tests based on the reading. You can develop true/false statements, multiple-choice questions, short answer, or essay questions. The teacher will collect the best test questions and include them in a comprehensive test on the reading selection.

ANALYSIS OF LITERATURE

Here are some suggested games and activities:

1. **A Book a Day.** Reading books often is one of the best ways to develop an understanding and appreciation for literature. The teacher and/or the students choose a book to read in one day. They rip the book apart into sections of 1-2 chapters. Students then divide into groups of two or four. They read their section, take notes on the key ideas, and discuss what they read. Then each group draws pictures and words based on the characters, setting, and events in the section. Finally, the teacher reviews the entire book with the class with each group explaining their section of the book.

2. **Poetry Gallery.** The teacher posts ten to twenty short to medium-length poems around the room. As the students enter the classroom, they are told to take out their notebooks and walk through the poetry gallery. As they read each poem, they would decide on the one they like the best and the one they dislike the most. They would copy these poems into their notebooks and answer these questions:

   What did you like or dislike about the poem? Describe what is happening in the poem. List a few examples of figurative language. Is there a theme or message the poem is trying to convey? Does this poem have any special meaning for your life? Why? With whom would you share this poem? Why?

3. **Anticipation Guide.** The teacher distributes a handout with a short excerpt from a work of literature. The author's name and the title of the work are omitted. The teacher or a student reads the selection aloud while the students follow along from the handout. For ten to fifteen minutes, the students write their impressions of the author and the work. They should include what we can learn about the author's values and beliefs from the selection. Students then share their impressions with the class. The teacher shares the name of the author and his/her background and the title of the work with the class. Students reread the selection and discuss if their impressions changed as a result of the additional information.

Variation. Students can read the same selection or a new selection and write their impressions of the aesthetic qualities of the work. These qualities include style, diction, figurative language, plot, character, tone, theme, and impact on the reader. They share their impressions and the teacher then provides additional information about the aesthetic qualities of the work.

4. **Class Newspaper.** This project can help you understand the setting, historical background, author's values and beliefs, and the various cultural influences on a work of literature. For example, after reading *The Outsiders*, students would create a daily city newspaper from the early 1960s. The articles would contain eyewitness accounts of events in the book as well as ads, town announcements, editorials, and other parts of a newspaper from that time. Students work together in groups to revise their articles and create a compete newspaper edition.

   Variation: Instead of a newspaper, students can create an early 1960's radio or television news show and include features and television commercials from that time. Students could also try a talk show format. The teacher and students can choose other works of literature for these types of projects as well.

5. **Point of View.** After reading a work of literature, students write a brief description of the narrator and explain how the story reflects his/her perspective of the characters and events. Identify also the point of view. Students then retell the story from another character's point of view. How does this new perspective change the story? Is it an advantage or disadvantage?

6. **Concept Cards.** You can use concept cards to help you retain key ideas, facts, and definitions about literature. Concept cards are usually 3 × 5 or 5 × 8 index cards. On the front of the card, write the concept or term you are learning. On the back of the card, write a definition with an example from literature. Review these cards to reinforce your understanding of literary analysis. You can also use concept cards to test your comprehension of fiction or nonfiction. Write a question on the front of the card and the answer on the back of the card. Then quiz yourself on the information you are trying to learn.

A
action, 91
allegory, 31
allusion, 31, 40
antagonist, 91, 99
antonym, 35
argument, 68
assumption, 68
audience, 77, 78, 79, 81
    awareness, 65, 66
author
    bias, 74
    credential, 71, 81
    credibility, 65
    opinion, 74, 81
    point of view, 94
    purpose, 67, 81
    style, 81
    tone, 95
    voice, 70
author purpose
    types of, 67

B
bank document, 110
bureaucracies, 107
bureaucracy, 107
business writing, 65

C
character
    influences, 92
    types of, 91, 92
character trait
    types of, 91
clarity, 65, 70
conflict, 90, 92
    types of, 96
consumer document
    types of, 108, 120
context clue, 28, 40
    signal words, 29
contract, 108

D
definition clue, 29
description, 91
detail, 45
dialogue, 91
dictionary entry
    origin of words, 35
    part of speech, 35
    pronunciation key, 35
directly stated main idea, 46, 57
driver's license, 109
dynamic character, 92

E
employee notices
    types of, 110
evidence, 68
expository writing, 73

F
fallacy, 68
figurative language
    types of, 31, 40
first person, 94
flat character, 92

G
generalization, 68
graphic, 107

I
identity form, 109
imagery, 27, 40
implied main idea, 46, 57
inference, 57
influence, 90
informal material, 111
informational material
    types of, 107, 110, 111
instruction manual, 108
Internet, 37, 39
irony, 28
    types of, 40
irony of situation, 33, 40

J
jingo, 74
job application, 117, 118

L
language, 27
    types of, 80
limited, 94
link, 36
literary devices
    types of, 28
literature, 88
    parts of, 87
loan document, 110

M
main idea, 57
    types of, 46
metaphor, 30, 40
    extended, 31
mood, 28, 40
motivation, 92

N

narration, 91
narrator, 91
narrator., 99
news
    feature, 37
    report, 37

O

omniscient, 94

P

paraphrase, 57
periodical, 36
personal document
    types of, 109, 111, 120
personal writing, 65
personification, 30, 40
persuasive writing, 73
plot
    parts of, 96
product information guide, 108
protagonist, 91, 99
public document
    types of, 110, 115, 120

R

reference source
    types of, 40
relationship, 92
round character, 92

S

satire, 28, 40
second person, 94
simile, 30, 40
social writing, 65
static character, 92
style, 65, 70
subordinate idea, 49
summarize, 57
symbolism, 28
synonym, 35
synthesize, 107, 120

T

tax form, 109
theme, 50, 51, 57
    types of, 99
third person, 94
tips
    analyzing argument, 68
    identifying facts and opinions, 76
    implied main idea, 47
    recognizing author's bias, 74
    stated main idea, 46
    themes, 51
    using keywords, 38
tone, 28, 34, 70
    types of, 73
tone,, 40
topic, 45

U

universal symbol, 32
universal theme, 89, 99

V

verbal irony, 33, 40
visual aid, 107
visual aids, 107

W

warranty, 108
workplace document
    types of, 111, 112, 113, 114
workplace manual, 110
writing
    types of, 65, 73